# Getting Top Math Scores on the Digital SAT®

## A Guide for Strong Students

by

Richard F. Corn

Richard Corn, LLC $\sqrt{\frac{rc}{l^2c}}$

# Getting Top Math Scores on the Digital SAT®

A Guide for Strong Students

by

Richard F. Corn

ISBN: 978-0-9985849-2-8



Errata or supplemental information for this book may be found at http://www.mathprepbooks.com/errata.com

Published by Richard Corn, LLC

**Richard Corn, LLC** $\sqrt{\frac{rc}{l^2c}}$

## Preface to the first edition

As of February, 2023 when this first edition was completed, the College Board had published eight practice tests. Four tests were linear (non-adaptive) practice tests to be taken with pencil and paper, and four were adaptive digital practice tests. Some of the questions in one set of tests could also be found in the other set of tests. More details about the digital SAT® and more practice materials are expected to be published over the summer of 2023. After that occurs, this book will be revised.

In the meantime, this book is useful to US students who participate in digital SAT® pilots, students who will be taking the PSAT/NMSQT in October 2023, students who are trying to decide between the digital SAT® versus the ACT®, and international students who are currently taking the digital SAT®.

# Table of Contents

# 1. Introduction

This book is designed for strong math students who have a reasonable expectation of scoring over 700 on the digital SAT®. If you have completed or almost completed honors Algebra 2 with at least a B or regular Algebra 2 with at least an A-, then this book should be a good fit for you. If you have not achieved those academic results in school, you will probably be better off with material that provides lots of practice on the full math syllabus for the digital SAT®. Khan Academy is a good resource for a comprehensive review of that type, see https://www.khanacademy.org/test-prep/digital-sat. This book addresses only a small part of the math content on the digital SAT® and does not offer lessons in math or step by step solutions to lots of practice questions.

Through years of preparing students for math questions on standardized tests, I observed that understanding what a question is asking you to do is often the biggest obstacle that even strong math students must work to overcome. Although students at every level face this challenge, students with more modest target scores can skip/guess those questions and still reach their targets. However, students with higher target scores cannot afford to skip/guess many questions, and they especially cannot afford to skip/guess questions that are rated as difficult (more on that later). This book is written for strong math students who want to improve their ability to recognize and answer questions more efficiently.

If you can recognize that a question follows a pattern, then you can understand what you need to do and solve it more easily. Intended for students who already know their math content, this book provides insight into how to answer questions that conform to recurring patterns, along with some traditional review of a few math topics that are often forgotten. Stronger math students are typically taking pre-calculus or calculus when they sit for the digital SAT®.

Each topic in the book contains a pattern description, tips for spotting the pattern, some sample questions with solutions, and an annotated list of some official practice questions where the pattern could be applied. The College Board has posted four linear pencil-and-paper practice tests and four adaptive digital tests.

> In this book references to official practice questions from the linear pencil-and-paper tests are in the form x.y.z, where x is the test number, y is the module number and z is the question number. The linear official practice paper-and-pencil tests can be found here https://satsuite.collegeboard.org/digital/digital-practice-preparation/practice-tests/linear

## Calculators

You may use a calculator throughout the math section of the digital SAT®. You may use your own calculator (remember to bring it with you!) or the on-screen calculator supplied by Desmos. You may use either calculator as you work through the math section, but don't forget that some questions are solved more easily by hand!

I find that some questions are easier to solve using a graphing calculator and others are easier to solve using the on-screen calculator. My advice is that if you are able to get some experience with the Desmos online calculator prior to the test, you will find that it was well worth the effort. In general, I find that if you want to work with a graph, you will be better off using the Desmos online calculator versus a graphing calculator. If you have no experience with the Desmos online graphing calculator, simply use your own calculator or solve by hand. You can check whether your particular calculator is allowed by visiting the College Board web site, see https://satsuite.collegeboard.org/digital/what-to-bring-do/calculator-policy .

Tips for solving questions with a graphing calculator and the online Desmos calculator appear throughout this book. Graphing calculator tips and screenshots are based on the TI-83/84 family of calculators. If you are using the TI-89 or the TI-Nspire calculators, your steps, keystrokes and screens will be different. The online calculator tips and screen shots are based on the test mode Desmos calculator, which is a version of their calculator used for online testing (see https://www.desmos.com/practice). The interface of the test mode version is similar to the full function Desmos calculator, but the test mode lacks advanced features. Although I have been careful to use basic features in the tips, it is possible that the version of Desmos used on the digital SAT® will be a little different from the Desmos test mode calculator used to illustrate this book.

## Module adaptive testing

Although there is only one math section on the test, it is delivered in two modules. Perhaps the biggest adjustment you will have to make for the digital SAT® is that although the statistical characteristics of the first module of the math section are fixed in advance and are the same for everyone, the second module of the math section is adaptive. Module adaptive testing means that the average difficulty of questions presented in the second module will depend on how well you did on the first module.

In the first module, the computer will present a collection of 22 questions on various topics and with various levels of difficulty. This module must be completed within 35 minutes and two questions will not count toward your score (these are experimental questions used only by the test makers and you will not know which questions they are). Although the questions may be different for each student taking the test in a particular location, all collections of the first module questions will be statistically equivalent with each other. They will have the same mixture of difficulty, the same mixture of topics, etc.

After the first module is completed, there will be a brief pause while the computer loads a second module of questions with the appropriate average level of difficulty. The better that you did in the first module, the higher the average level of difficulty will be in the second module. Just like the first module, the second module will have a collection of 22 questions on various topics and various levels of difficulty. However, the average level of difficulty will depend on how well you did on the first module. The second module must be completed within 35 minutes and two questions will not count toward your score (these are experimental questions used only by the test makers and you will not know which questions they are). Your final math score will reflect a combination of results from both modules.

The illustration below is meant to give you an idea of how module adaptive testing could work. The diagram shows a first module collection followed by three possible second module collections. Each second module will contain easy, medium and hard questions, but the average level of difficulty (indicated by the capitalized label in the diagram) will differ. As stated earlier, all first module collections are statistically equivalent. The second module collections are statistically equivalent within each average level of difficulty.

Unlike what is illustrated below, second module levels of difficulty could result in overlapping final math scores because the final math score is based on results from first and second modules combined. In order to illustrate how module adaptive testing works, we are trying to keep the diagram simple (and we do not know how big the overlaps will be). We also do not know how many average levels of difficulty there will be in the second modules. There are three levels in the illustration, but as of February 2023 it looks like the digital SAT® may have only two levels of difficulty for second modules.

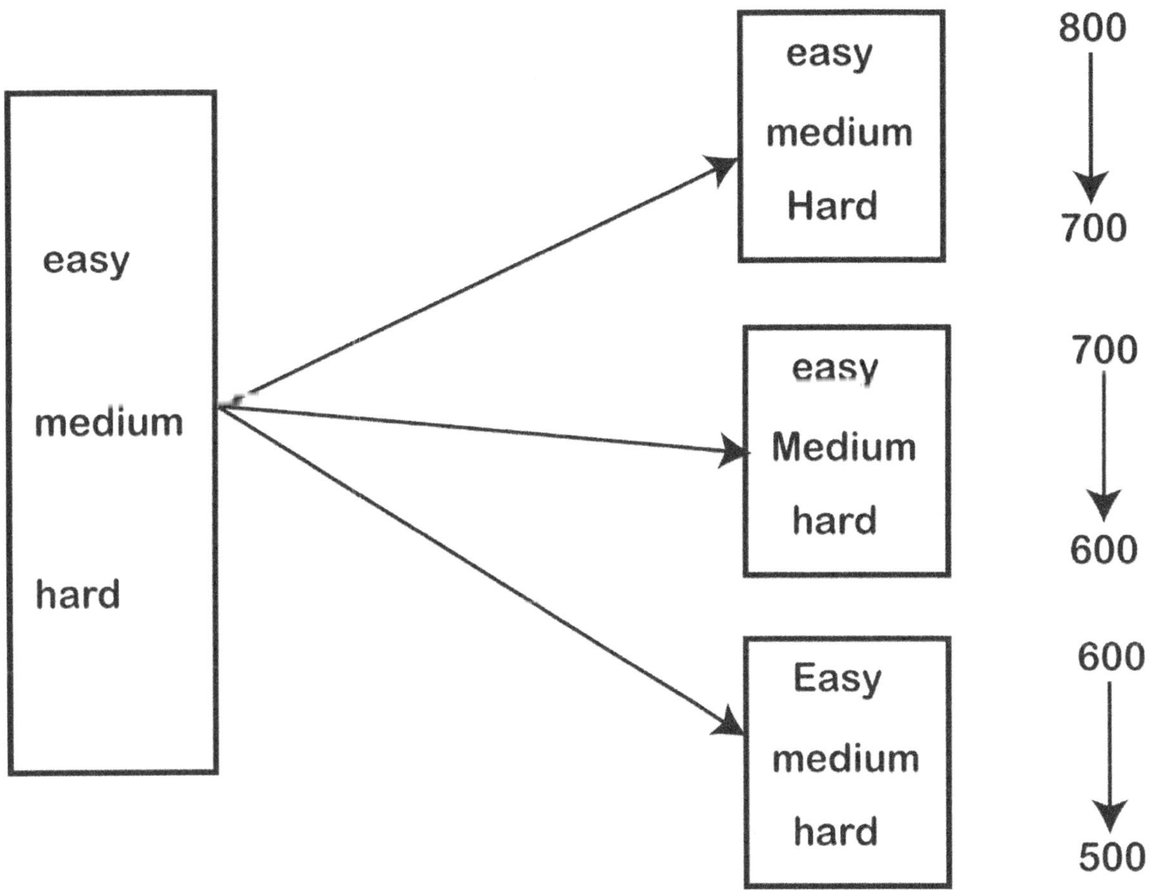


You can get some insight into how you did in the first module by what questions you see in the second module. If you are seeing questions that are challenging it means that you did well in the first module, but if you see questions that are mostly easy it may mean you did not do as well in the first module as you should have. Remember that, as shown in the illustration, second module questions will vary in difficulty regardless of the average level of difficulty. Therefore, you should not panic if the first several questions in the second module are easy because you could still be in a second module with a higher average level of difficulty.

Once you are placed into a second module level based on your answers to the first module, your highest possible score and lowest possible score are probably limited. For example, if you bomb the first module it may not be possible to score above 600 overall, even if you answer every question in the second module correctly. On the other hand, if you aced the first module you have a shot at an 800, but only if you do well in the second module. If you ace the first module and bomb the second module you might not be able to score above 700.

## Questions are weighted

In school, math questions are weighted differently. Some questions might have more weight because they are more difficult and others might have more weight because they have multiple parts. So, you should be comfortable with the idea of a test where questions are weighted differently.

As in school, if you want a very high score on the digital SAT® you will have to reach the more heavily weighted questions that are expected to appear toward the end of each module and solve them correctly. Note that we also expect that questions in the first module will be dispersed broadly across levels of difficulty, whereas in the second module questions are expected to be dispersed more narrowly (fewer easy questions). Also as in school, if you make too many silly mistakes in the easier questions, you might be disappointed with your score even though you answered the more difficult questions correctly.

## Putting all this together

What will it take to get a high score? The first thing you must do is make very few mistakes in the first module so that you are placed into an appropriate second module that reflects your ability. If you make too many mistakes in the first module you might be placed in a second module that is unacceptably easy, meaning that it limits your overall math score to something below your target. The next thing you must do is answer most of the questions in the second module correctly. If you did very well in the first module, most of the questions you see in the second module should be harder. The hardest questions are expected to appear toward the end of each module. If you want to get a very high score, those must be solved correctly because they carry more weight.

Just like in school, pacing is an issue here. A handy feature of the digital SAT® is the online timer, which shows you how much time is remaining in the module. The advice below is based on what we have seen so far in the published practice tests.

- In the first module, time is not likely to be an issue. <u>You should not rush to finish.</u> My recommendation is to move at a comfortable pace, making sure that you <u>do not make silly mistakes</u>. That is the key to the first module. Slow down so that you can check your work as you go. You should finish the first 11 problems with fewer than 25 minutes remaining on the online timer (this is an average of about one minute per problem in the first half of the module). A little less than 25 minutes remaining would give you a little over 2 minutes per question for the remaining 11 questions and that should be more than enough time. Remember that this recommendation is just a starting point. Keep track of your pace as you work through the first module of your practice tests and make adjustments for the next practice test if needed.

- Your approach should be to assume that you have been placed into a more difficult second module. If that turns out to not be the case you will have to take the test again anyway. You should have a sense of whether you were placed into a more difficult second module fairly early on.

  The pacing for the second module is quite different. Sometimes the second module will be a slow build of difficulty level throughout the module, but other times the second module may have several very challenging and/or time-consuming questions toward the end. My recommendation is to assume a challenging finish and hope for the slow build. Move through the first half of the second module "with all deliberate speed," which is a handy military phrase that means go as fast as you can without making mistakes. That would mean 25-30 minutes remaining on the online timer at the end of the first half of the module, which is an average of 30-40 seconds per question for the first half. Having 25-30 minutes remaining gives you an average of 2-3 minutes per question for the second half of the second module which should allow you to weather a potential storm at the end. Remember that this recommendation is just a starting point. Keep track of your pace as you work through your practice tests and make adjustments for the next practice test if needed.

A statistical breakdown of the math content on the digital SAT® is:

- About 30% of the problems on the math section will be word problems. Most of these will contain fewer than 50 words.
- About 75% of the questions will be multiple choice, the rest will be student produced response (grid-in) questions.
- About 70% of the questions will focus on algebra, whereas the balance will be on data analysis/probability (15%) or geometry/trigonometry (15%).

## Recommendations for taking the test

Make sure that the proctor gives you scratch paper before the test starts. You are not supposed to bring your own scratch paper because the proctor would have to check every side of every page that you (and other students) bring. Bring your own pencils or pens (whatever you are most comfortable using with scratch paper).

Unless you are expert at using the online Desmos graphing calculator, bring your own calculator. Or to put this more bluntly: always bring a calculator.

When answering a math problem, use scratch paper to set up the problem. Then finish the problem on the scratch paper or on either calculator, whichever is most comfortable for you. Many honors classes in school do not make much use of calculators. Don't forget that some questions are solved more easily by hand!

## Answering multiple choice questions

As you are working through the questions in a section, follow these steps for multiple choice questions:

- Read the question carefully. If you know how to solve for the answer using the math you learned in school, go ahead.
- If you cannot solve using the math you learned in school, determine whether the question conforms to one of the patterns in this book. If so, then solve accordingly.
- If you cannot solve the question using the math you learned in school or using a pattern, try to eliminate answer choices in multiple choice questions. Keep in mind that diagrams are drawn to scale unless otherwise indicated, which means you can trust your eye.
- Random Guessing: For multiple choice questions, randomly select a remaining answer choice. Do not leave a multiple choice question blank.
- If you have run out of time, randomly guess answer choices for the remaining multiple choice questions.
- If you finish with time remaining, go back to questions where you were uncertain of your answer. Consider using the Mark for Review tool (look for the flag icon on the context menu on the top right of the screen) to mark questions where you guessed or are uncertain of your answer. Or you could make a note on your scratch paper as you work through the test.

## Answering student-produced response questions

The digital SAT® contains multiple-choice questions and student-produced response questions (also called "grid-in questions" or "type-in-the-answer questions" or "free response questions"). For this test, the best name for these questions would be type-in-the-answer because that is literally what you do.

For student produced response questions, if you have no idea about how to answer the question leave it blank, do not spend extra time guessing. If you have calculated an answer but do not have confidence in it, enter the answer anyway. Come back to it later if you have time.

A question may have more than one correct answer. If that should happen, you should enter the value of any of the correct answers (enter only one answer).

On student-produced response questions the computer will display a box into which you could enter at most six characters. These characters can consist of digits (0-9), a decimal point, a slash (/) that can be used for fractions, or a negative sign (-). No other characters will be accepted.

If your answer is a negative value then enter the negative sign first, leaving you five remaining characters to use. If your answer is a positive value, you may only enter up to five characters. Do not be concerned about right or left justification. If your positive answer is above 99999 then your answer is wrong; if your negative answer is below -99999 then your answer is wrong

What if your answer consists of more than five characters (plus a negative sign for negative answers)?

- If your answer is a fraction that cannot fit in five characters (including the slash but not including the negative sign), use your calculator to convert it to decimal and follow the instruction below.
- If your answer is a decimal number that cannot fit in five characters (including the decimal point but not including the negative sign) you must round it to five characters or truncate it.

For example, suppose your answer is -128/11. The fraction is seven characters long, including the negative sign and slash, too big to fit. When converted to decimal you find that it is $-11.\overline{63}$. To make it fit in six characters you could truncate (which would result in -11.63) or you could round it to -11.64 (because -11.6363 rounds to -11.64).

## Recommendations for preparing for the test

Start preparing at least two months before your first live attempt at the test. Set up a schedule for yourself so that you are doing several hours of preparation each week. I suggest starting with the linear pencil and paper tests in order to get a sense of the content and question styles of the test. Those must be followed by online practice tests using the Bluebook™ software. It is important to become familiar with the online tools for marking questions, etc. If you have only been practicing math sections, before the test date make sure to take a full length adaptive Bluebook™ practice test (with verbal and math sections).

After each practice test that you take, determine the reasons for any errors you made. If you did not know how to solve the problem, use this book to determine whether the question follows one of the patterns. If you did not know the math needed to solve the problem, you could refer to a math textbook, a comprehensive review book, or online resources such as Khan Academy or YouTube to learn the math on your own.

During your preparations, try to work through every step by hand if practical (for example, there is no need to do complicated long division by hand). Then try to check it by using your calculator or Desmos. If you are not familiar with the Desmos online calculator I recommend that you experiment with it well before your test date. It is intuitive and there are lots of help videos on YouTube. You will find that learning how to graph with Desmos is worth the effort.

## Reader feedback, errors and supplements

I enjoy hearing from readers, so please feel free to send an email to mathprepbooks@gmail.com . Despite a thorough editing of the material in this book it is inevitable that mistakes were missed and I apologize for them in advance. If you find an error, please send an email. Errors found in the book will be posted promptly at mathprepbooks.com/errata.html, along with links to any supplemental materials that were written after its publication.

# 2. Widely applicable patterns

These patterns have broad applicability on the digital SAT®, meaning that they can be applied to any question for any math topic at any level of difficulty. These patterns are not connected to math content per se, but are general tips for how you should approach the test. One of the main challenges is to train yourself to take this test very differently from how you would take a math test in school.

## 2.1 Use information contained in the answer choices

**Why this is awkward:** Most students see few if any multiple choice questions on math tests in school. A common mistake is for students to behave as if they were in school: they will read a problem and then solve it without considering the answer choices. However, if you understand the problem but do not see how to use math to solve it, then you should consider using the information in the answer choices. Sometimes using the answer choices is easier or faster than using the traditional math approaches you learned in school. The two methods discussed here are backsolving and substitution. They can be used separately or in combination.

**How to spot the pattern:** If you are working on a multiple-choice problem and you understand what the problem is asking you to do but you do not see how to use math to solve the problem, consider backsolving and/or substitution instead of trying to solve as if you were in school.

**Pattern examples**

<table>
<tr>
<td>

**Backsolving example:**

Which ordered pair satisfies the inequality $y<-3x-4$ ?

A) (3, 1)<br>
B) (-3,1)<br>
C) (1,3)<br>
D) (1,-3)

</td>
<td>

To solve using math, you would sketch the equation of the line and shade the solution region. Next you would check each answer choice to decide whether or not the point lies in the solution region.

If you forgot how to solve inequalities you could backsolve. Plug each answer choice back into the inequality to test whether the inequality is satisfied.

A) $1<-9-4$<br>
B) $1<9-4$<br>
C) $3<-3-4$<br>
D) $-3<-3-4$<br>
SOLUTION: B

Notice that backsolving is faster than solving this problem using math.

</td>
</tr>
</table>

| | | |
|---|---|---|
| **Calculator Tip:** | Enter an equation into the *Y*= equation editor as shown on the top. Change this to a less-than inequality by placing the cursor over the box-slash area, then press enter. Toggle until you find the less-than icon and press enter (as shown in the middle).<br><br>Press the *graph* key. The solution region is shaded. | 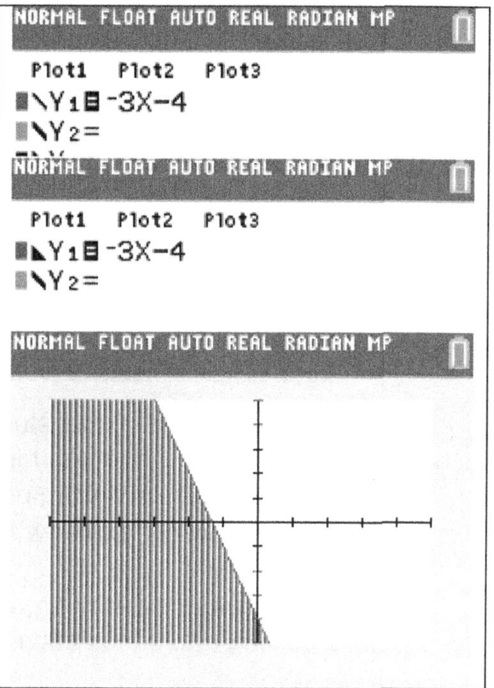 |

| | | |
|---|---|---|
| **Online Calculator Tip** | Enter the inequality in the equation window and press enter. The solution region is shaded. | 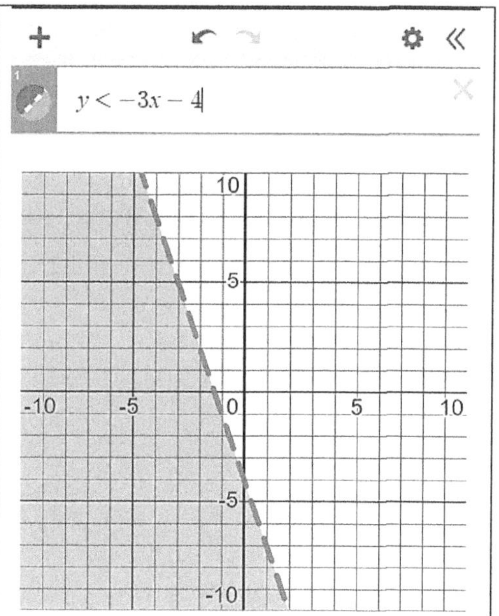 |

**Another backsolving example:**

What value of $x$ satisfies the inequality $2|x-3|>8$ ?

A) -2

B) -1

C) 0

D) 1

To solve this using the math you learned in school, you would follow these steps:

$$2|x-3|>8,\quad |x-3|>4$$
$$x-3<-4 \text{ or } x-3>4$$
$$x<-1 \text{ or } x>7$$

SOLUTION: A

If you forgot how to solve greater than absolute value inequalities, you could backsolve:

A) $2|-2-3|=2\cdot5=10$

B) $2|-1-3|=2\cdot4=8$

C) $2|0-3|=2\cdot3=6$

D) $2|1-3|=2\cdot2=4$

**Calculator Tip:**

Enter two equations into the *Y*= equation editor as shown. Note that the absolute value function is found by pressing *math-num-abs(* . After the equations are entered, press *graph*. If you cannot see the graph very well, adjust the window by pressing *window*.

The solution region is wherever the absolute value curve lies above the horizontal line $y=8$.

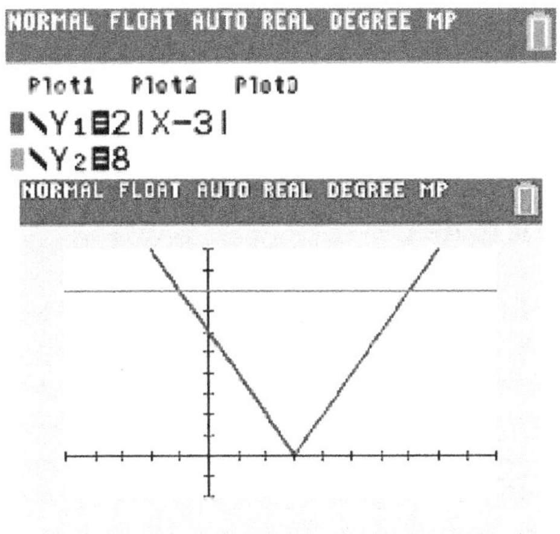

| Online Calculator Tip | Enter the two equations (the absolute value can be found on the numerical portion of the online keyboard).<br><br>The solution region is wherever the absolute value curve is above the horizonal line $y = 8$ | 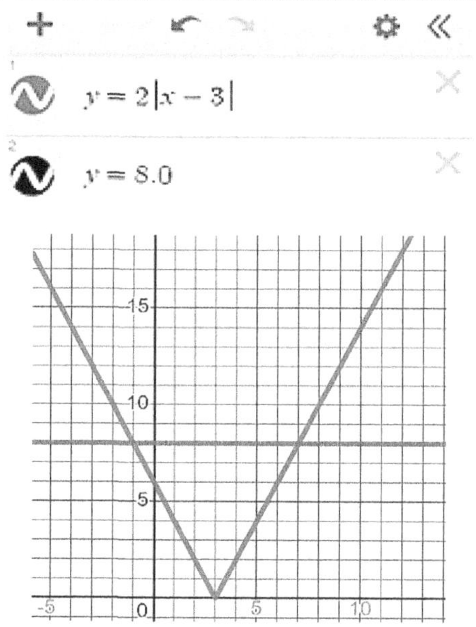 |
|---|---|---|

| **Substitution example:**<br><br>A store clerk was asked to markup the price of a shirt by 20%. Instead, the clerk marked the price down by 20%. By what percentage does the price now have to be increased in order to be correct?<br><br>A) 30%<br>B) 33%<br>C) 40%<br>D) 50% | Using algebra, you would let $x$ be the original price of the shirt. Then the shirt was incorrectly reduced to $.8x$ and now the price must be increased from $.8x$ to $1.2x$, That represents an increase of $0.4x$, or 50% of the incorrect price.<br><br>SOLUTION: D<br><br>Instead of algebra just use substitution. Let the original price be 100. It was incorrectly reduced to 80 but now must be increased to 120, which represents a 50% increase of the incorrect price. |
|---|---|

**Substitution and backsolving example:**

Which of the following is equivalent to $\dfrac{3x^4+x^3-2x^2+x+8}{x^2+2x+1}$?

A) $3x^2-5x-15+\dfrac{36x-23}{x^2+2x+1}$

B) $3x^2-5x-5+\dfrac{16x+3}{x^2+2x+1}$

C) $3x^2-5x+5+\dfrac{16x+13}{x^2+2x+1}$

D) $3x^2-5x+5+\dfrac{-4x+3}{x^2+2x+1}$

If you forgot how to do polynomial long division. you can use substitution and backsolving instead.

The problem does not place restrictions on $x$ so we will pick a convenient value, $x=0$.

When $x=0$, $\dfrac{3x^4+x^3-2x^2+x+8}{x^2+2x+1}=\dfrac{8}{1}=8$.

Plugging zero into the answer choices gives us A) as -38, B) as -2, C) as 18 and D) as 8.

SOLUTION: D

Notice that using substitution and backsolving could be faster (and less error prone) than long division.

NOTE: When using substitution and backsolving, sometimes you can get two answer choices that are both correct. When that happens choose a different value of $x$ and then re-calculate those two answer choices to break the tie.

**Comments on some official problems:**

1.1.26 Notice that two equations are the same so you do not need to check them both. Notice that answer choices (C) and (D) have $x=r$. So that would mean $y=(-2r+7)/3$ which eliminates (C) and (D). Notice that answer choice (B) has $y=r$. So that would mean that $x=(-3r+7)/2$ , hooray!

1.2.18 This is a great place for substitution. Let the 2015 population be 100. Then the 2016 populations must be 107. Solve the equation $107=k100$.

1.2.19 If you are uncertain about laws of exponents This is a substitution/backsolving opportunity. Let a=2. Then $2^{\frac{11}{12}}=1.8877$. Plug 2 into each of the answer choices.

1.2.26 This is an excellent backsolving example. First put the equations into slope-intercept form. From the first system, solve to find that $\dfrac{-a}{b}=\dfrac{7}{5}$. Use that to backsolve the answer choices until you find that a pair of equations that are perpendicular. Answer B works because $\dfrac{-a}{2b}=\left(\dfrac{1}{2}\right)\left(\dfrac{7}{5}\right)=\dfrac{7}{10}$.

2.1.22 Start backsolving for when x=-6. f(-6)-3=0-3=-3. That eliminates (A) and (D). Next try x=-5. f(-5)-3=0-3=-3. That eliminates (C).

2.2.15 Let x=10. Then $\frac{8x(x-7)-3(x-7)}{2x-14}=\frac{231}{6}=28.5$ Now plug 10 into the answer choices until you find one that equals 28.5.

## 2.2 Interim steps as answer choices

**Why this is awkward:** Sometimes you are so relieved to see how to calculate an answer that you think you are finished when, in fact, what you calculated (correctly) was the result for an interim step. This pattern is awkward because in school you almost are always solving for $x$, whereas on the SAT you are often asked to solve for a different variable or you are asked to solve for the value of an expression containing $x$, such as $2x-1$.

**How to spot the pattern:** Notice exactly what the problem is asking you to solve for. If it is a word problem and you are setting up your own equation and your own variables, you can avoid mistakes of this type by letting $x$ represent what the problem is asking for. If the problem already has variables assigned and/or an equation or equations to solve, be careful when you find the value for one of the variables. Make certain that the value of that variable is what you are asked to find.

**Pattern examples**

|  |  |
|---|---|
| A piggy bank contains pennies, nickels and dimes. If there are three times as many pennies as nickels and there are two times as many nickels as dimes, how many nickels are there if the value of the contents of the piggy bank is \$1.30 ?<br><br>A) 5<br>B) 10<br>C) 20<br>D) 30 | To be extra careful we set $p$ to represent pennies, $x$ to represent nickels, and $d$ to represent dimes. We choose $x$ to represent nickels because that is what the problem asks us to solve. Now we can write the equations:<br>$p=3x$<br>$x=2d$<br>$.01p+.05x+.10d=1.30$<br>Next, we re-write the last equation in terms of $x$ and then solve:<br>$.01(3x)+.05x+.10(x/2)=1.3$<br>$0.13x=1.3,\ x=10$<br>SOLUTION: B. Notice that there are answer choices for pennies and dimes. |

|  |  |
|---|---|
| If $\frac{8^x}{128^x}=\frac{1}{256}$ then $2^x$ must be equal to:<br><br>A) -2<br>B) 2<br>C) 4<br>D) 8 | First re-write the equation in terms of base 2:<br>$\frac{2^{3x}}{2^{7x}}=\frac{2^0}{2^8}$<br>Then apply laws of exponents to solve:<br>$2^{-4x}=2^{-8},\ -4x=-8,\ x=2$<br>$2^x=2^2=4$<br><br>SOLUTION: C. Notice that $x=2$ is an interim solution to the problem. |

| | | |
|---|---|---|
| **Calculator Tip:** | Enter two equations into the *Y=* equation editor as shown. Adjust the window so that you have a decent view of the intersection. Then press *2^nd^-trace* to get into the calc menu. Choose *5: intersect* and follow the dialog. You should have a solution that looks like the next screen shot. The coordinates of the intersection are at the bottom of the screen. So $x = 2$ and $2^x = 4$. | 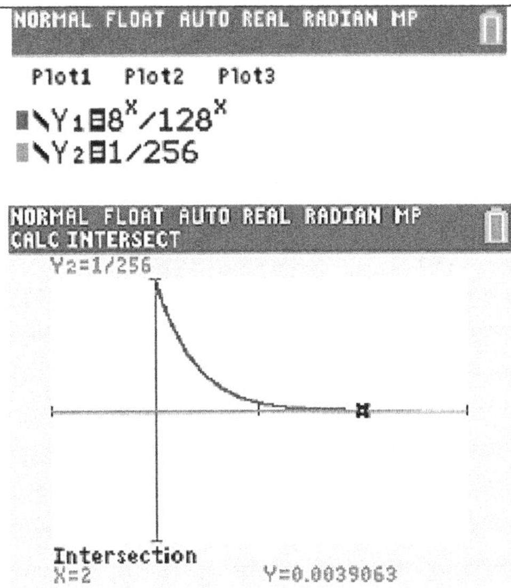 |
| **Online Calculator Tip** | Enter the two equations (you can click on the keyboard icon or just type the equations in).<br><br>Place your mouse on either curve and click. You will notice a dot appear over the point of intersection. Place your mouse on the dot and the (x,y) coordinates of the dot will appear on the screen. So $x = 2$ and $2^x = 4$ | 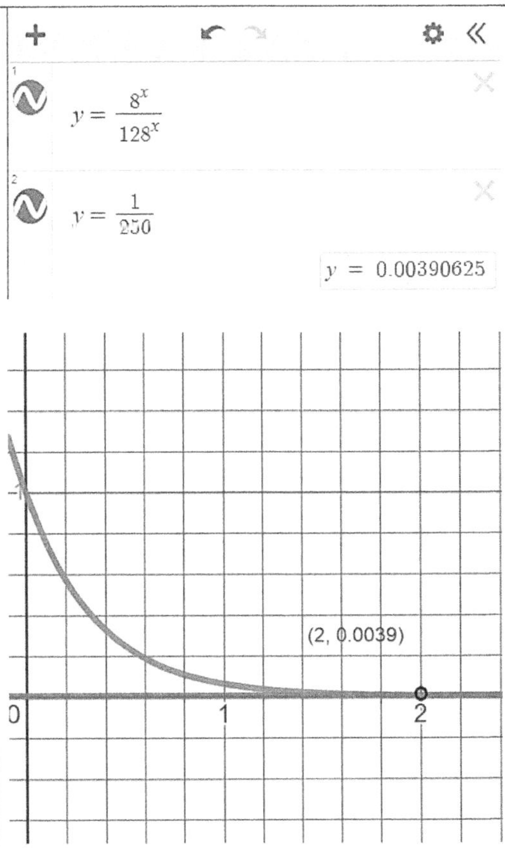 |

| The solution to the equation $x^2 + 8x + 3 = 0$ takes the form of $x = a \pm \sqrt{b}$. What is the value of $a + b$?<br><br>A) -8<br>B) -4<br>C) 9<br>D) 17 | Using the quadratic formula we get<br>$x = \frac{-8 \pm \sqrt{64 - 4(1)(3)}}{2} = \frac{-8 \pm \sqrt{52}}{2}$<br>$= \frac{-8 \pm 2\sqrt{13}}{2} = -4 \pm \sqrt{13}$<br><br>SOLUTION: C.<br>Notice that $a = -4$ and $b = 13$ are interim solutions to the problem.<br><br>Notice that a calculator may not be very useful here. |
|---|---|

**Comments on some official problems:**

2.2.16 You may be so relieved to find that the slope of line p is -9 and you see -9 is the first answer choice. You choose the shiny object rather than remember that what is being asked is the slope of line r.

3.2.12 The steps are: $48 = 8\sqrt{x}$, $6 = \sqrt{x}$, $36 = x$ Notice that 6 is one of the answer choices.

3.2.18 You solve for x and find that it is 7 and eagerly move on rather than solve for y.

4.1.9 When you see that Angle Q is congruent to Angle T you may jump to answer 18 degrees without noticing that the problems asks for Angle S.

4.2.16 Set angle x and angle y equal to each other to find that k is equal to 21. You choose answer choice B without realizing that the question is asking for z, not k.

## 2.3 Multiple choice questions where a diagram is drawn to scale

**Why this is awkward:** What makes this pattern awkward is that you almost never see diagrams drawn to scale in school. Some students erroneously assume that a diagram they see on a test in school or in a textbook is drawn to scale and end up making mistakes. On the digital SAT®, a diagram drawn to scale presents a fantastic guessing opportunity that students often do not exploit.

**How to spot the pattern:** Unless indicated otherwise, all diagrams on the digital math SAT® are drawn to scale. If an angle looks like it is 30 degrees, you can assume that it is approximately 30 degrees. If one line segment looks twice as long as another line segment, you can assume that it is approximately twice as long. If you are told in the question that the measure of an angle is 45 degrees then it is exactly 45 degrees in the diagram.

**Pattern example**

| In the diagram below, what is the measure (in degrees of the angles marked X ?<br>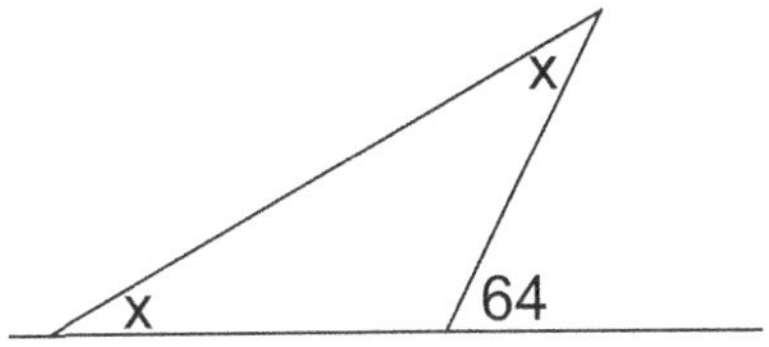<br><br>A) 12<br>B) 22<br>C) 32<br>D) 52 | This is easy to solve. We know the obtuse angle of the triangle must be $180-64=116$. Therefore $x+x=180-116,\ x=32$. Or just remember the exterior angle theorem.<br><br>If this diagram were to appear on the digital math SAT® exactly as shown, we could take an excellent guess because the diagram is drawn to scale (there is no indication otherwise). Compare the angles marked x with the angle marked 64. The angles marked x seem to be about half size of the angle marked 64, guess 32. Or at least eliminate A and D and then guess.<br><br>SOLUTION: C |
|---|---|

This pattern is well suited to cherry picking problems when time is running out. As you scan the remaining problems, look for a multiple choice question with a diagram. If you cannot solve it quickly, take a good guess by eliminating answer choices.

**Comments on some official problems:**

Other than charts or graphs, none of the diagrams on the linear practice tests are drawn to scale.

## 2.4 Roman numeral questions

**Why this is awkward:** Students may be unfamiliar with roman numeral questions, which are just a special kind of multiple choice question.

**How to spot the pattern:** Due to their special format, roman numeral questions are easy to spot. Their structure lends itself to a good guessing opportunity even if you cannot solve the question completely.

**Pattern example**

| | |
|---|---|
| Two sides of a triangle are each 5 units long. Which of the following statements MUST BE true?<br>I. At least two of the angles of the triangle are congruent.<br>II. The third side of the triangle must be less than 10 units long.<br>III. The triangle is equilateral.<br><br>A) I only<br>B) II only<br>C) I and II only<br>D) I and III only | Approach this as three true/false questions.<br><br>I. True. If two sides of a triangle are congruent then the triangle is isosceles and at least two angles are congruent.<br><br>II. True. The triangle inequality states that any side of a triangle must be less than the sum of the other two sides.<br><br>III. False. The triangle could be equilateral, but it might not be (all we know for sure is that it is isosceles).<br><br>If you are able to determine true or false for I, II, and III with complete certainty, then you just pick C. Suppose you are only certain that I is true. Then eliminate B and guess. If you are only certain than II is true, eliminate A and D and then guess. If you are certain that III is false then eliminate D and guess.<br><br>SOLUTION: C |

**Comments on some official problems:** There are no Roman Numeral questions in the linear practice tests, however there are some in the adaptive Bluebook™ practice tests.

# 3. Algebraic patterns

These patterns are used in algebra problems or algebraic calculations generally.

## 3.1 Unit conversion (dimensional analysis)

**Why this is awkward:** Most students think they are done with unit conversions when they finish chemistry. Unfortunately, unit conversions appear again on the digital SAT® as a type of word problem.

**How to spot the pattern:** Unit conversions can be difficult to spot because you often do not realize that you have a unit conversion until you have started to work the problem. If you see a conversion ratio somewhere in the problem, like (1cup is equivalent to 8 fluid ounces), that is a good indication that you are working on a unit conversion problem (lucky you!). If you do not see a conversion ratio, start by reasoning your way to a solution. Once you realize you are doing unit conversion you can re-start the problem as one big unit conversion or you can pick up where you left off in order to finish the remaining steps.

### Pattern examples

| | |
|---|---|
| If angle X is $\frac{\pi}{3}$ radians smaller than angle Y, and the measure of angle X is $\frac{3\pi}{4}$ radians, what is the measure of angle Y in degrees? | It is clear that this is a unit conversion problem, from radians to degrees. But first we need the measure of angle Y in radians.<br>$\frac{3\pi}{4} = m\angle Y - \frac{\pi}{3},\ m\angle Y = \frac{3\pi}{4} + \frac{\pi}{3} = \frac{9\pi + 4\pi}{12} = \frac{13\pi}{12}$<br>To get the measure of angle Y in degrees we use<br>$\left(\frac{13\pi}{12}\right)\text{radians}\left(\frac{180\text{ degrees}}{\pi\text{ radians}}\right) = 195\text{ degrees}$<br>Notice that you are expected to know the ratio for converting between degrees and radians. |

| | | |
|---|---|---|
| Calculator Tip: | To convert $(13\pi/12)$ radians to degrees, press the mode key to make sure your calculator is in degree mode.<br>Next, enter $(13\pi/12)$ into the calculator followed by 2nd-apps-3 to indicate that your entry is in radian mode. Then press enter. | 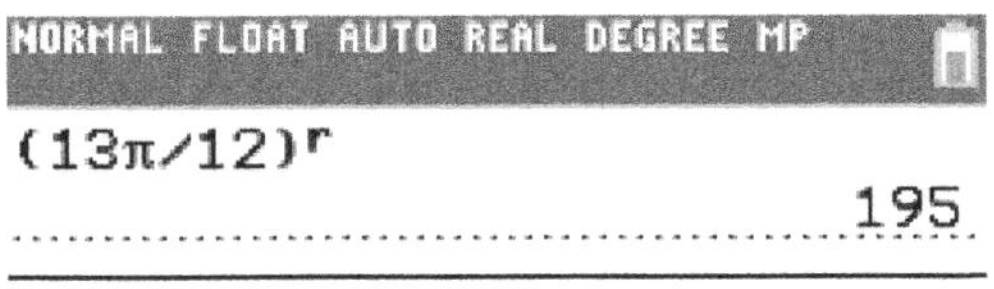 |

| Online Calculator Tip | Enter the same equation as shown above in the solution by hand. You should see the answer in the bottom right portion of the equation box.<br><br>You will also be shown a graph but you have to zoom out to see it. | 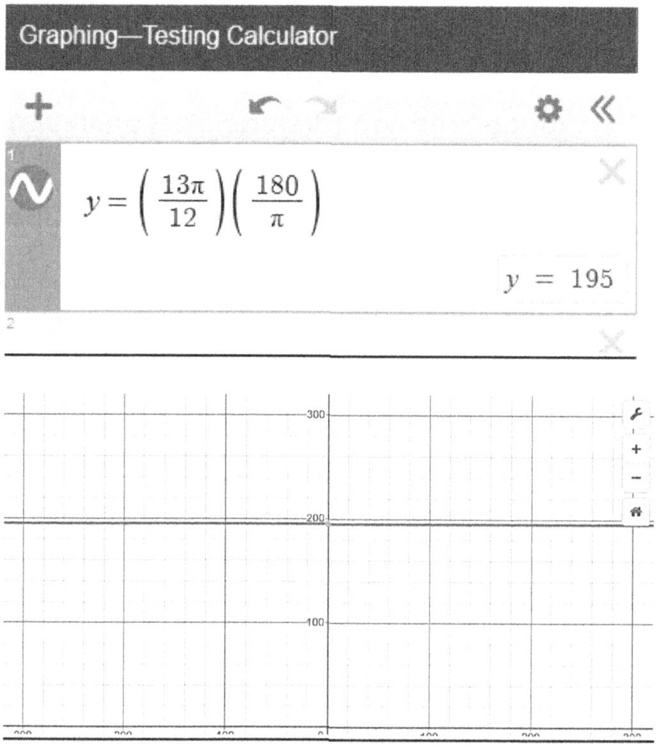 |
|---|---|---|

| | |
|---|---|
| If a car's fuel efficiency is 20 miles per gallon of gasoline and gas costs $4 per gallon, how much will it cost to travel 100 miles? | Start setting up the equation by establishing the target unit, which is dollars because we are asked how much travel will cost. <br> $= \text{dollars}$ <br> Next start with the first piece of information given which is 20 miles per gallon. <br> $\left(\frac{20 \text{ miles}}{1 \text{ gallon}}\right) = \text{dollars}$ <br> Because the target is dollars we realize that we must eliminate miles and gallons. We can eliminate gallons by inserting the cost per gallon. <br> $\left(\frac{20 \text{ miles}}{1 \text{ gallon}}\right)\left(\frac{4 \text{ dollars}}{1 \text{ gallon}}\right) = \text{dollars}$ <br> Although we are happy to see dollars in the numerator for both sides of the equation, we must also eliminate gallons. This is done by flipping the first term. <br> $\left(\frac{1 \text{ gallon}}{20 \text{ miles}}\right)\left(\frac{4 \text{ dollars}}{1 \text{ gallon}}\right) = \text{dollars}$ <br> Now we are left with dollars per mile on the left side. We can eliminate miles by inserting 100 miles to the equation. <br> $\left(\frac{1 \text{ gallon}}{20 \text{ miles}}\right)\left(\frac{4 \text{ dollars}}{1 \text{ gallon}}\right)(100 \text{ miles}) = \frac{(4)(100)}{20} \text{ dollars}$ <br> Now we are done because we have dollars on the left and dollars on the right! It will cost $20 to travel 100 miles. |

**Comments on some official problems:**

1.1.23 Set up the target $4.36 \text{ mi}^2 = yd^2$

Introduce the other information $4.36 \text{ mi}^2 \left(\frac{1760\, yd^2}{1\, mi^2}\right) = yd^2$

2.1.26 Set up the target $\left(\frac{1 \text{ gallon}}{220 \text{ ft}^2}\right) = \text{p gallon}$

Introduce the other information $2\left(\frac{1 \text{ gallon}}{220 \text{ ft}^2}\right)\left(w \text{ ft}^2\right) = \text{p gallon}$

2.2.23 Set up the target $11{,}863{,}808 \text{ yd}^2 = \text{mi}^2$

Introduce the conversion factor $(11{,}863{,}808\, yd^2)\left(\frac{1 \text{ mi}^2}{1760\ yd^2}\right) = \text{mi}^2$

## 3.2 Solve for one variable in terms of other variables (literal equations)

**Why this is awkward:** In school we are accustomed to solving for the variable named $x$ and we are accustomed to solving for a particular value, as in $x = 1.34$. In this pattern we are not always solving for a variable named $x$ and we are not trying to determine the value of the variable of interest. Instead, we are trying to express the variable of interest in terms of other variables and constants. This is done by isolating the variable of interest from all of the other variables and constants. These problems are called "literal equations."

**How to spot the pattern:** Look for multiple choice questions where (1) the answer choices are equations with only one variable on one side of each equation or (2) the answer choices are expressions that involve all but one of the variables. That one isolated or missing variable is the variable of interest. When answer choices are equations, solve the problem by isolating the variable of interest. If the answer choices are expressions where one variable is missing, solve for the variable that is not shown.

**Pattern examples**

| | |
|---|---|
| The study of the physics of whistles makes use of the Strouhal number ($S$) which can be found by using the equation $S = \frac{fL}{U}$, where $f$ is the flow speed of the whistle, $L$ is the diameter of the cylinder of the whistle, and $U$ is the speed of the airflow over the whistle. The speed of the airflow over the whistle can be expressed as:<br><br>A) $U = SfL$<br><br>B) $U = \frac{fL}{S}$<br><br>C) $U = \frac{fS}{L}$<br><br>D) $U = \frac{SL}{f}$ | To solve, you typically would read the question carefully, thinking that it is an awkward word problem. Then you might spend even more time on figuring out what the question is asking you to do.<br><br>Remember that there can be important information in the answer choices. A quick look at the equation inside the problem and the answer choices tells you that you are being asked to solve the equation for $U$. This can be seen by noticing that the answer choices are forms of the equation where only $U$ is on one side of the equation.<br><br>SOLUTION: B<br><br>$S = \frac{fL}{U}$<br>$SU = fL$<br>$U = \frac{fL}{S}$ |

| The value of $\pi$ was discovered by the Babylonians and the Egyptians at least 4,000 years ago. Which of the expressions below did they probably use to calculate $\pi$?<br><br>A) $\dfrac{\text{area of a circle}}{\text{its radius}}$<br><br>B) $\dfrac{\text{area of a circle}}{\text{its diameter}}$<br><br>C) $\dfrac{\text{circumference of a circle}}{\text{its radius}}$<br><br>D) $\dfrac{\text{circumference of a circle}}{\text{its diameter}}$ | Even if you are annoyed by the history lesson and have no idea how $\pi$ was calculated thousands of years ago, a quick glance at the answer choices will do the trick. Each answer choice is a formula that might have been used to calculate $\pi$. What makes this difficult to decipher is that answer choices are expressions rather than equations where $\pi$ is on only one side of the equation.<br><br>This problem also requires you to know the equations for the area of a circle and the circumference of a circle.<br><br>SOLUTION: D<br><br>Write the equation of the area of the circle and solve for $\pi$.<br><br>$A = \pi r^2$, $\pi = \dfrac{A}{r^2}$. Unfortunately this does not match any answer choice.<br><br>Write the equation of the circumference and solve for $\pi$<br><br>$C = 2\pi r$, $\pi = \dfrac{C}{2r}$ |
|---|---|

**Comments on some official problems:**

2.2.10 Isolate the x, expressing it in terms of the other variables.

4.1.19 Isolate the radical and then square both sides.

## 3.3 Interpreting the slope or y-intercept of linear models

**Why this is awkward:** It can be surprisingly difficult to interpret the meaning of slope or the meaning of the y-intercept of a linear relation in the context of a word problem or even a graph or table. You must start by understanding what is happening in the problem and then try to solve.

**How to spot the pattern:** Make sure that you are looking at a linear relation. If the question asks for a rate, a constant rate or an average rate then it is asking about slope. If the x-variable represents time and the question asks for a starting value of $y$, then the question is asking about the y-intercept. For multiple choice problems, eliminate answers that have the wrong slope or the wrong y-intercept.

**Pattern examples**

| | |
|---|---|
| After walking up a hill at a constant speed for 10 minutes, Julia reached an elevation of 20 feet. If the hill had an initial elevation of 5 feet, what is the total time that Julia will have walked at the same constant speed when she reaches an elevation of 100 ft?<br><br>A) 33<br>B) 33<br>C) 50<br>D) 63 | The phrase "constant speed" tells us that the elevation is a linear function of time, or $e = mt + b$ where $e$ is elevation, $m$ is speed, $t$ is elapsed time, and $b$ is the initial elevation. Plugging in the point (10, 20) into the equation $e = mt + 5$ (the initial elevation is 5 ft) gives us $20 = 10m + 5$. Solving for the speed $m$ gives us $\frac{20-5}{10} = 1.5$. We solve for the total time walked by evaluating the equation $100 = 1.5t + 5$ which gives us 63.3 minutes.<br><br>SOLUTION: D |

The scatter plot below illustrates the density of particles over a period of time.

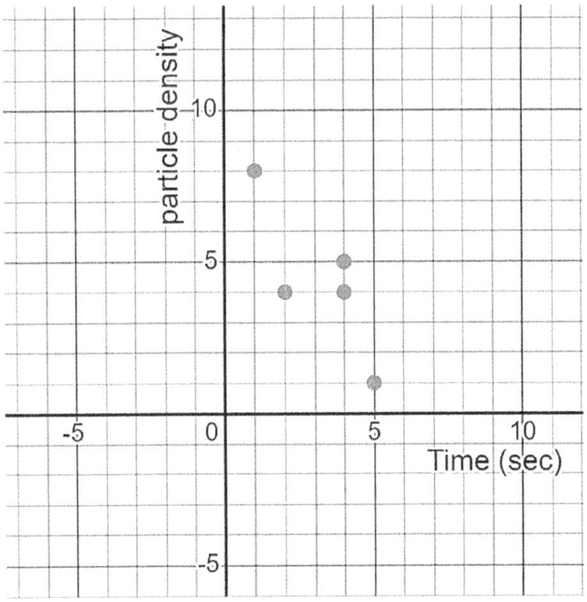


Which of the following conclusions could <u>not</u> be made based on the data?

A) Particle density decreases at a rate of 1.2 particles per second.

B) Thc initial particle density was approximately 5 units.

C) After more than 5 seconds elapse there are few, if any, particles remaining.

D) The initial particle density was approximately 9 units.

Check each answer choice very carefully. If you were to sketch a line of best fit, the equation would be something like $y = -1.2x + 9$.

SOLUTION: D

**Comments on some official problems:**

3.1.9 Any time you see the word "rate" it means slope. If they wanted the y-intercept it would be the electrician's onetime fee.

3.1.12 Notice that they are giving you two points, (0, 41) and (1, 40). Use those points to find the slope equals -1, which eliminates (C) and (D). You know the y-intercept is 41, not 0.

3.1.15 You have to recognize that 17 is the starting amount and the y-intercept. Then you recognize that the rate is the slope, giving the equation $y = -\frac{1}{4}t + 17$.

## 3.4 Linear Systems: One solution, no solution, infinite solutions

**Why this is awkward:** When you see a system of linear equations you might be tempted to solve the system using substitution or elimination because that is what you have been doing in school for years. But if the answer choices are one solution, no solution, or infinite solutions then the best approach does not involve solving.

**How to spot the pattern:** There are three good ways to spot this pattern:

- Notice that you have a system of two equations and both equations are linear.
- Scan the problem and answer choices for the phrases: "one solution," "no solution" or "infinite solutions."
- Notice that you are not being asked to solve the system.

First, write both equations in slope-intercept form. If the problem asks for one solution then the two slopes cannot equal each other (intersecting lines). If the problem asks for no solution then the two slopes must equal each other and the y-intercepts cannot equal each other (parallel lines). If the problem asks for infinite solutions then the two slopes must equal each other <u>and</u> the y-intercepts must equal each other (overlapping lines).

**Pattern examples**

<table>
<tr>
<td>In the system of equations below, what value of $a$ would produce an infinite number of solutions?<br><br>$ax + 5y = 10$<br>$2x - 3y = 6$<br><br>A) 2<br>B) -10/3<br>C) any value of $a$ would produce an infinite number of solutions<br>D) no value of $a$ would produce an infinite number of solutions</td>
<td>The slopes must be equal and the y-intercepts must be equal. Start by writing the equations in slope-intercept form:<br><br>$5y = -ax + 10$<br>$3y = 2x - 6$<br><br>$y = (-a/5)x + 2$<br>$y = (2/3)x - 2$<br><br>Setting the slopes equal to each other will cause $a$ to be -10/3. However, the y-intercepts would not match.<br><br>SOLUTION: D</td>
</tr>
</table>

<table>
<tr>
<td>Calculator Tip:</td>
<td>We will use the calculator to backsolve this. First, we store the first value for $a$ into one of the variables. I am choosing variable V. Enter 2 and press the store button sto-> followed by apha-6. You should see the first screen on the right.<br><br>Next enter the slope-intercept equations into the Y= equation editor. When entering the equations use the variable V instead of $a$ by pressing alpha-6. You should have equations that look like the second screen on the right.<br><br>Press the 2nd-graph key to see the graph. We see the that the lines intersect.<br><br>Continue to backsolve by changing the value of V by using the store button. You will see that the lines are parallel when V is -10/3 but do not overlap.</td>
<td>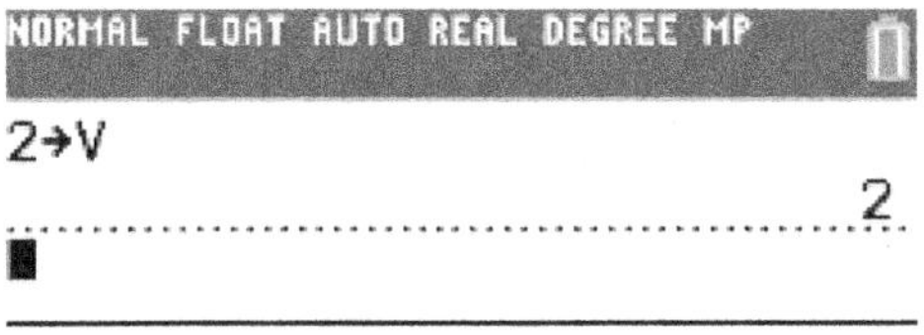

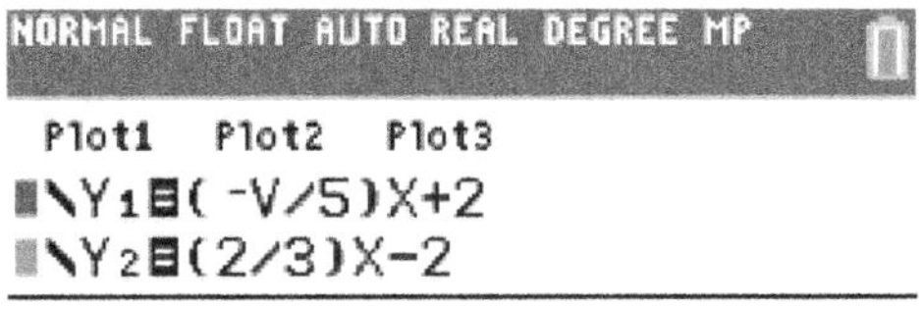

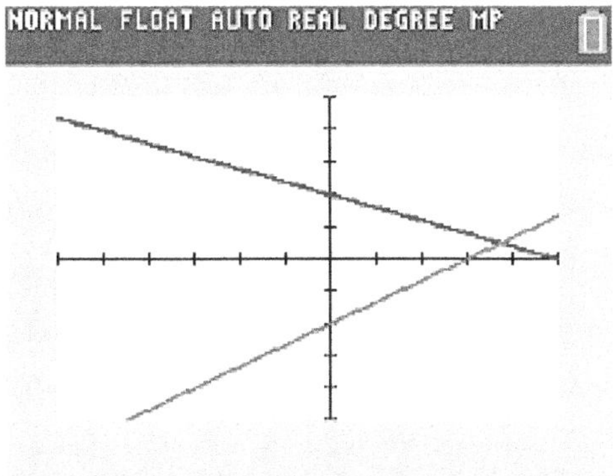

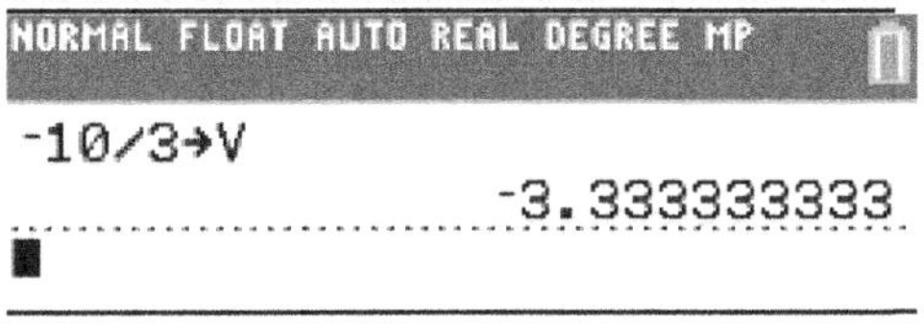
</td>
</tr>
</table>

| Online Calculator Tip | Enter the first equation and accept the slider for $a$. Enter the second equation.<br><br>Move the slider around. No matter where the slider is placed, the graph shows that the lines never overlap. | 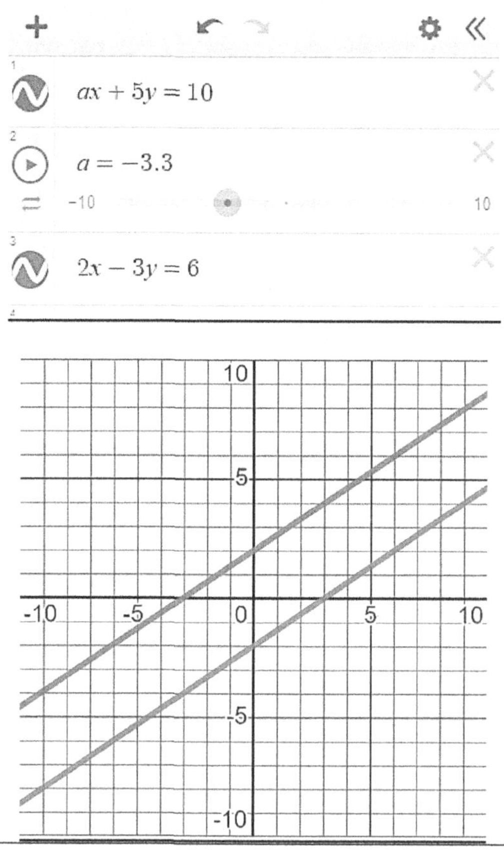 |
|---|---|---|

**Comments on some official problems:**

2.2.19 At first this does not look like a system of equations, but it is. The tipoff is the phrase "no solution." The second equation must have the same slope of 6 and a y-intercept not equal to 18.

3.2.27 First re-write each equation in slope-intercept form. Then equate the slopes $\frac{1}{12} = \frac{1}{2p}$

Notice that the y-intercepts are different.

## 3.5 A system of a line and a parabola, or a system of two parabolas

**Why this is awkward:** In school you rarely see a system of equations where one equation is a line and the other equation is a parabola. If you are solving by hand, you would set the two equations equal to each other and go from there. You are less likely to see a system of two parabolas but if you did you would set the two equations equal to each other and go from there.

These problems are best solved graphically, but you may have to make multiple adjustments to the viewing window to be certain that you are not missing intersections

**How to spot the pattern:** There are two good ways to spot this pattern:

- Notice that you have a system of two equations where one equation is linear and the other equation is quadratic, OR you have a system of two equations where both are quadratic.
- Scan the problem and answer choices for the phrases: "one solution," "two solutions" or "no solution." These may not appear, but if you notice them you are a step ahead.

**Pattern examples – a line and a parabola**

| How many solutions exist for the system of equations below? | Start by equating the expressions for $y$, then solve. |
|---|---|
| $y = 2x^2 - 4x - 6$<br>$y = 5x - 13$ | $2x^2 - 4x - 6 = 5x - 13$<br>$2x^2 - 9x + 7 = 0$<br>$(2x - 7)(x - 1) = 0$ |
| A) no solution<br>B) one solution<br>C) two solutions<br>D) infinite solutions | We can see that there are two solutions at x=7/2 and x=1.<br><br>SOLUTION: C |

| Calculator Tip: | Enter the equations into the Y– equation editor.<br><br>Press the *graph* key to see that the graphs intersect at two points. | 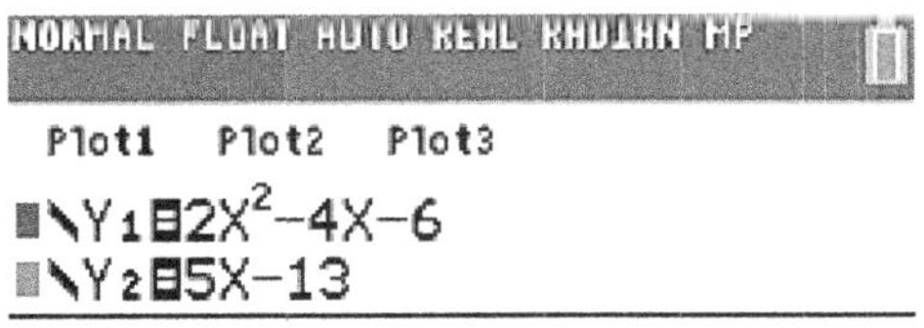 <br>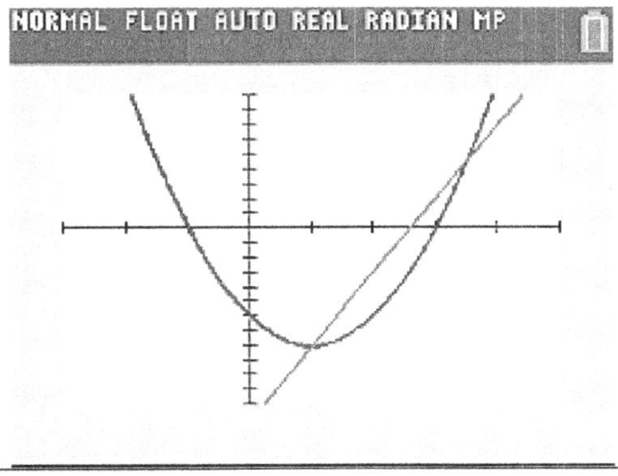  |
|---|---|---|

| Online Calculator Tip | Enter the two equations.<br><br>The graph shows that there are two points of intersection. | 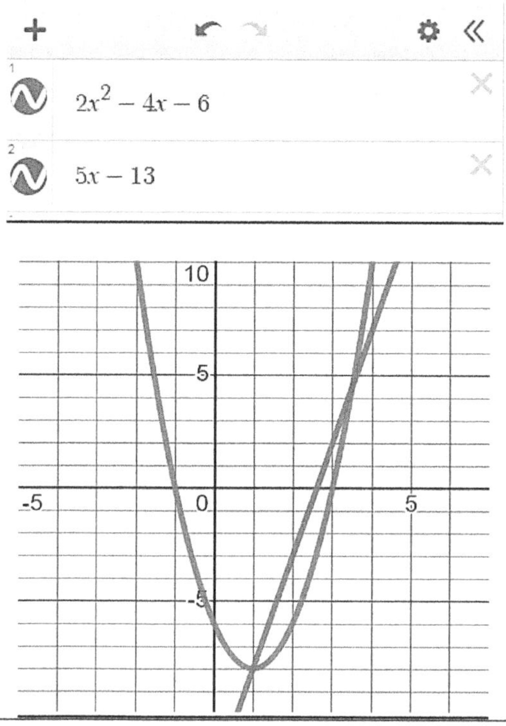<br> |
|---|---|---|

| If the line $y = k$ and the curve $-3x^2 + 5x + 1$ intersect at only one point, what is the value of $k$ ?<br><br>A) 5/6<br>B) 19/6<br>C) 21/36<br>D) 111/36 | Notice that $y = k$ is the equation of a horizontal line and that $-3x^2 + 5x + 1$ is the equation of a parabola opening down. If only one intersection is allowed, it must occur at the vertex of the parabola, which is located at<br><br>$x = \frac{-b}{2a} = \frac{5}{6}$ Therefore<br>$y = -3\left(\frac{5}{6}\right)^2 + 5\left(\frac{5}{6}\right) + 1 = \frac{111}{36}$<br><br>SOLUTION: D |
|---|---|

| Calculator Tip: | Enter the quadratic equation into the Y= equation editor, then press the graph key.<br><br>Enter 2nd-trace-4 to start the dialog to find the maximum value. Follow the prompts. You should see a screen similar to that shown. | 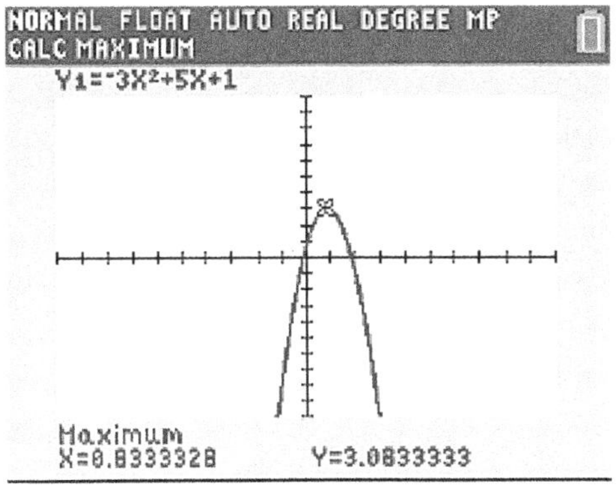 |
|---|---|---|



| Online Calculator Tip | Enter the quadratic equation. The enter a function with a slider.<br><br>Adjust the slider until the line goes through the vertex of the parabola. The answer is that $k$ is approximately 3.1 | 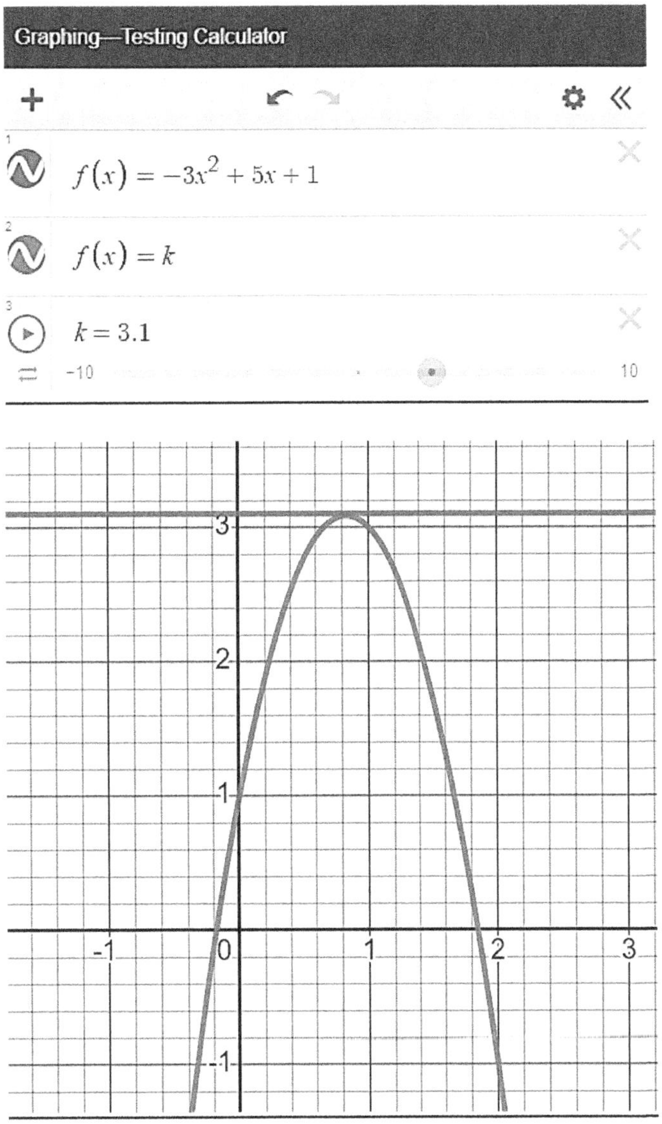 |
|---|---|---|

### Pattern examples –two parabolas

| How many solutions exist for the system of equations below? | Start by equating the expressions for $y$, then solve. |
|---|---|
| $y = .5x^2$<br>$y = x^2 + 8x + 5$<br><br>A) no solution<br>B) one solution<br>C) two solutions<br>D) infinite solutions | $x^2 + 8x + 5 = .5x^2$<br>$.5x^2 + 8x + 5 = 0$<br><br>Use the discriminant to test for the number of solutions:<br><br>$D = b^2 - 4ac = 64 - 4(.5)(5) = 54$<br><br>Because the discriminant is positive, we know that there are two solutions.<br><br>SOLUTION: C<br><br>The calculator notes below illustrate how the calculator could steer you wrong. **It is not known whether something like this would occur on the test, but it might.** |

| Calculator Tip: | Enter the equations into the Y= equation editor.<br><br>Press the *graph* key to see the graph. If your window is in the standard position or in most other reasonable settings, you might conclude that there is only one solution.<br><br>However, if you adjust the window upwards you will eventually see that the parabolas intersect again at (-15.3, 117.8).<br><br>This illustrates the danger of relying on the graphing calculator. | 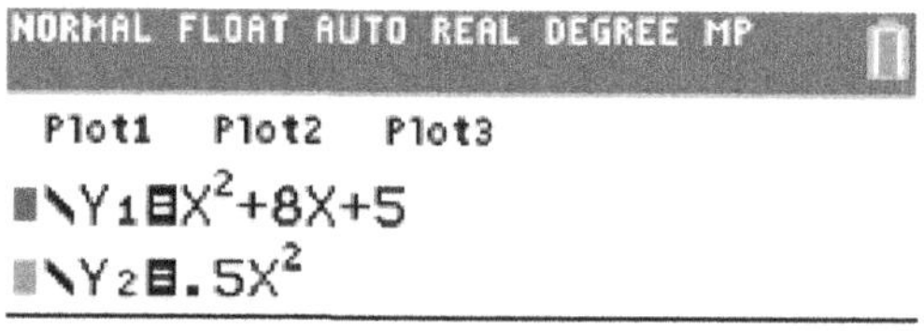<br>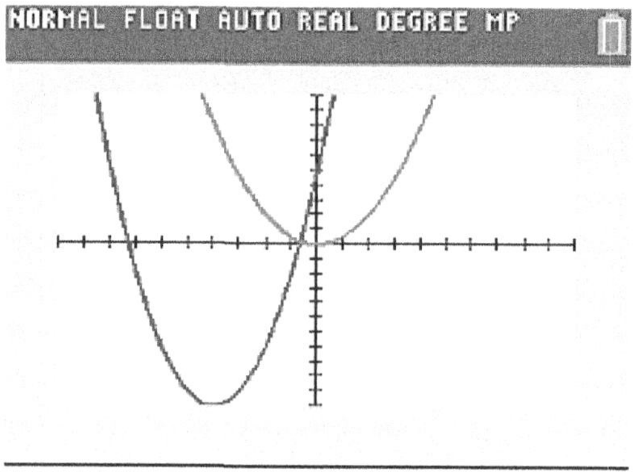 |
|---|---|---|

| Online Calculator Tip | Enter the two equations.<br><br>If your window is in the default position or in most other reasonable positions, you might conclude that there is only one solution.<br><br>However, if you adjust the window upwards you will eventually see that the parabolas intersect again at (-15.3, 117.8).<br><br>This illustrates the danger of relying on the graphing calculator. | 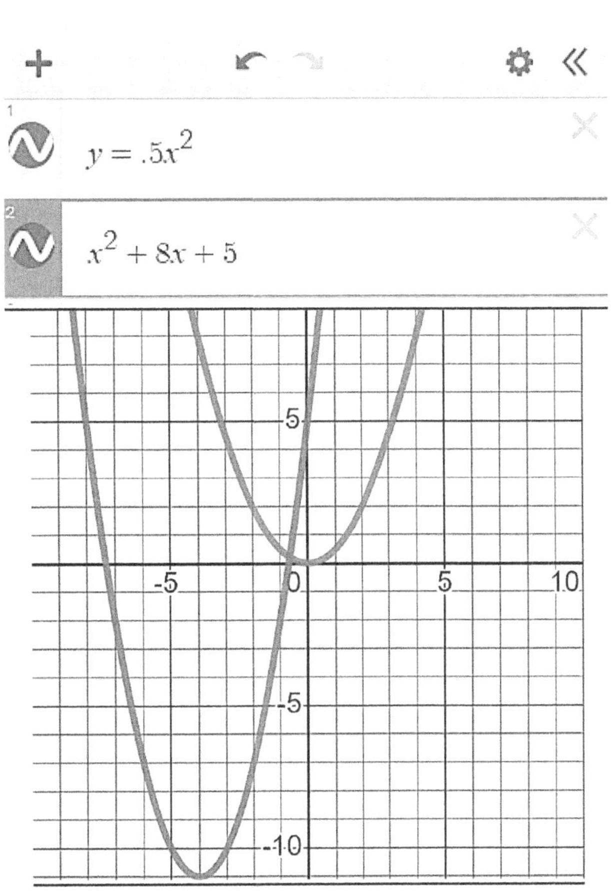 |
|---|---|---|



**Comments on some official problems:**

1.1.25 Because the line is horizontal and there is only one point of intersection, and that point must be the vertex of the parabola. Use -b/2a to find the x-coordinate of the vertex. Then plug that into the equation of the parabola.

4.1.24 Set the two equations equal to each other and then bring all terms to the same side, giving you $2x^2 - 24x + (64 - a) = 0$. Because there is one point of intersection set the discriminant equal to zero, which will give you $576 - 4(2)(64 - a) = 0$.

Note that this question is set up perfectly for Desmos. Enter both equations and activate the slider for $a$. Manipulate the slider until the line intersects with the parabola at only one point (the slider will be at approximately -8.1). Click on what looks like the point of intersection. Notice that the x-coordinate of the intersection is approximately equal to 6.

## 3.6 Discriminant Pattern

**Why this is awkward:** When you see a quadratic equation set equal to zero (or a quadratic equation that could easily be set equal to zero), you might be tempted to solve the equation by factoring or by quadratic formula because that is what you have been doing in school for years. But if the question does not actually ask for the solutions, the discriminant can save you time.

The discriminant is the expression under the square root sign in the quadratic formula, $b^2 - 4ac$. As shown below, the sign of the discriminant can tell you whether there is one real solution, two real solutions, or no real solutions (two imaginary solutions).

Two real solutions

$b^2 - 4ac > 0$

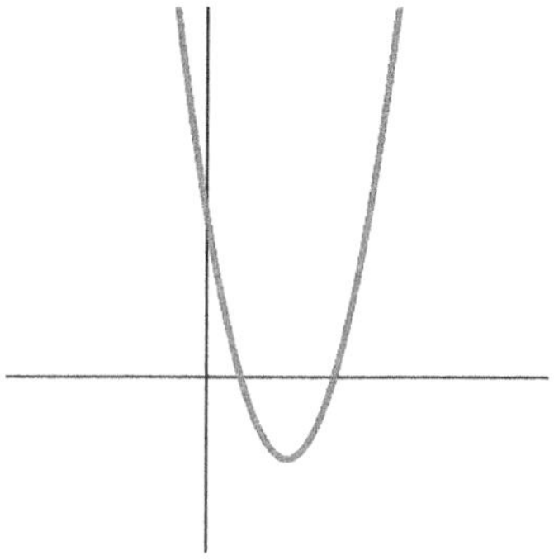

One real solution

$b^2 - 4ac = 0$

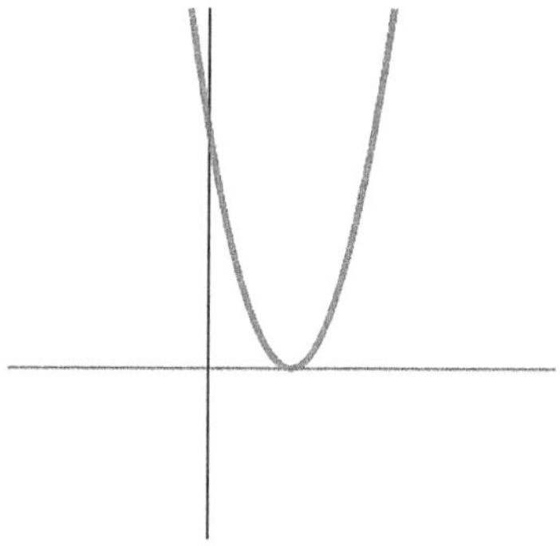

No real solutions or two imaginary solutions

$b^2 - 4ac < 0$

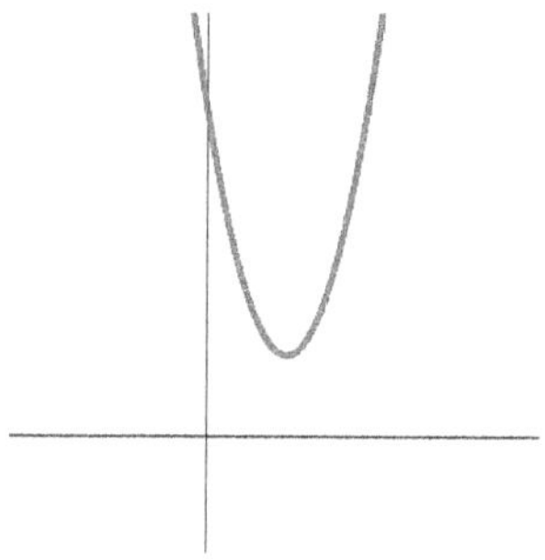

**How to spot the pattern:** There are three good ways to spot this pattern:

- Notice that you have a quadratic equation set equal to zero or could easily be set equal to zero.
- Scan the problem and answer choices for the phrases: "one solution," "one real solution, "no solution," "no real solution," or "two imaginary solutions."
- Notice that you are not being asked to solve.

**Pattern examples**

<table>
<tr>
<td>Suspect that $f(x)=-2x^2+3x-2$ and $g(x)=f\left(x-2\right)$. Which of the following would be the outcome to the equation $g(x)=0$ ?<br><br>A) no real solutions<br>B) one real solution<br>C) two real solutions<br>D) two rational solutions</td>
<td>$g(x)=-2(x-2)^2+3(x-2)-2=-2x^2+11x-16$<br>$b^2-4ac=121-4(-2)(-16)=121-128=-7$<br><br>Because the discriminant is negative, there are no real solutions, just two imaginary ones.<br><br>SOLUTION: A</td>
</tr>
</table>

<table>
<tr>
<td>Calculator Tip:</td>
<td>Enter the equation $-2(x-2)^2+3(x-2)-2$ into the Y= equation editor.<br><br>Press the graph key to see the graph. Because the graph has no x-intercepts, there are no real solutions.</td>
<td>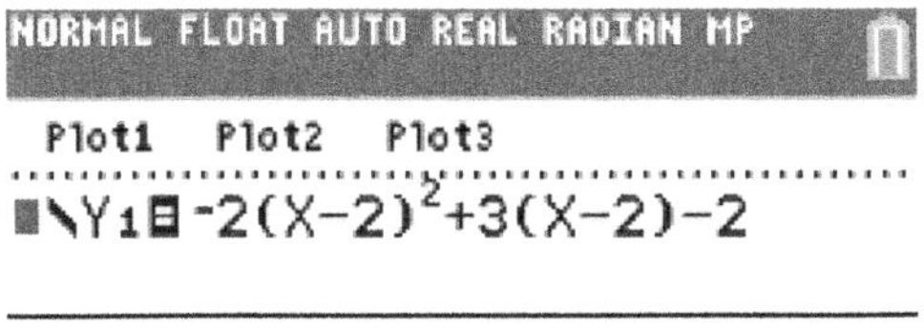<br>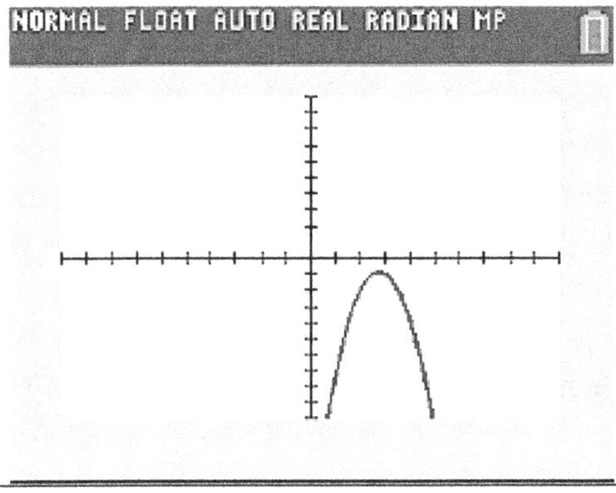</td>
</tr>
</table>

Online Calculator Tip

Enter the equation $-2(x-2)^2+3(x-2)-2$.

Because the graph has no x-intercepts, there are no real solutions.

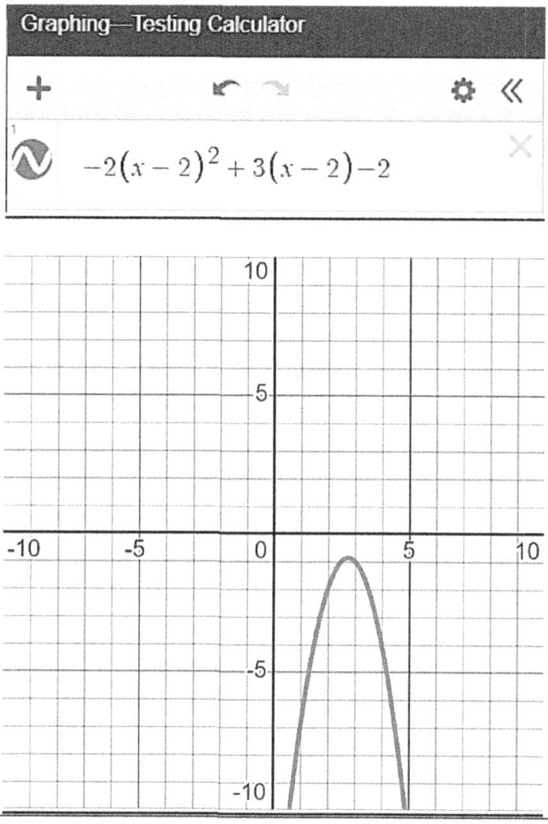


If the difference between the area of a circle and the circumference of that same circle is equal to $k$, which of the intervals below would <u>not</u> contain the value of $k$ ?

A) $k<-\pi$

B) $k\ge-\pi$

C) $-\pi\le k\le 0$

D) $k>0$

$\pi r^2-2\pi r=k,\quad \pi r^2-2\pi r-k=0$

discriminant: $(-2\pi)^2-4\pi(-k)<0$

$4\pi^2+4\pi k<0,\quad 4\pi(\pi+k)<0,\quad \pi+k<0.$ k<-$\pi$

SOLUTION: A

| Online Calculator Tip | Enter the equation<br>$\pi r^2 - 2\pi r - k$<br><br>Create a slider for $k$.<br><br>If you set $k = -\pi$ you will see from the graph that there is one value of $(r = 1)$ where the equation has one x-intercept. (shown on the right).<br><br>If you set $k$ to a value below $-\pi$ , you will see that the parabola has no x-intercept.<br><br>If you set $k$ to a value above $-\pi$ you will see that the parabola has two x-intercepts. | 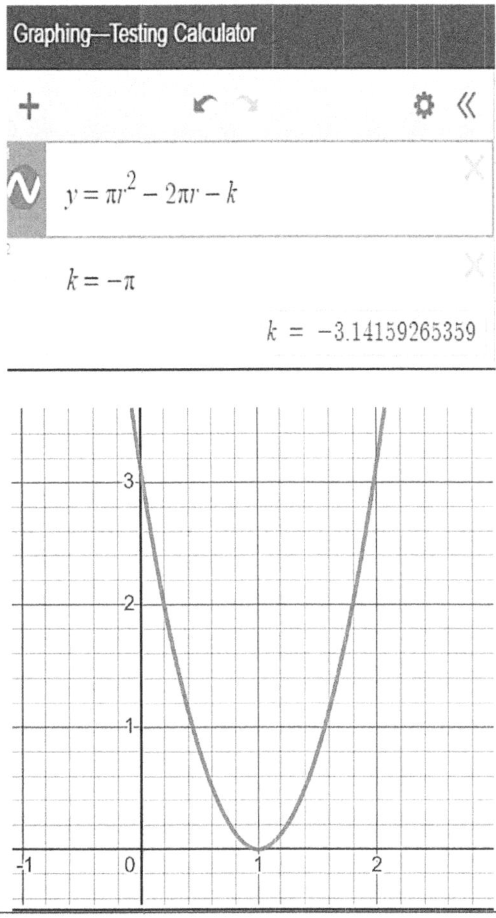<br> |
|---|---|---|

**Comments on some official problems:**

1.2.27 The tip off here is “no real solutions.” Start with $1156 - 4(1)c < 0$

2.1.21 The tip off here is “one distinct real solution.” Start with $-1.5 = x^2 + 8x + a$ and move over the -1.5. Set the discriminant equal to zero $64 - 4(a + 1.5) = 0$.

2.2.22 The tip off here is “distinct real solutions.” Start with $100 - 4(5)(16)$ to find that the discriminant is negative.

3.2.26 The tip off here is “exactly one solution.” Start with $30^2 - 4(-9)c = 0$.

4.2.21 The tip off here is “no real solutions.” Start with $b^2 - 4(-1)(-676) < 0$ and find that $-52 < b < 52$

## 3.7 Linear models vs. exponential models

**Why this is awkward:** You rarely have to think of these two models at the same time because you are taught linear models one year and then exponential models another year. However, if you think way back to Algebra 1, you might remember that a linear model occurs when evenly spread x-values produce y-values that either increase or decrease by the same amount (a common difference). If you think back to Algebra 2, you might remember that an exponential model occurs when evenly spread x-values produce y-values that either increase or decrease at the same rate (a common ratio). To jog your memory further, remember that you also studied arithmetic sequences (which are linear) and geometric sequences (which are exponential). Most important to remember is that in the exponential model, $a$ is the starting amount and $b$ is the base. Exponential growth at the rate of $r$ has this base: $b=1+r$ and exponential decay at the rate of $r$ has this base: $b=1-r$.

**How to spot the pattern:** There are three good ways to spot this pattern:

- Look at the answer choices! If you see that two of them are linear equations $y=mx+b$ and the other two are exponential equations $y=ab^x$ then you are in this pattern.
- If you are not given x-values that are evenly spread then invent a few. Then calculate the y-values corresponding to those x-values. If the y-values increase or decrease by the same amount then the relation is linear. If the y-values increase or decrease at the same rate then the relation is exponential.
- If the problem is about populations, biology or money, you are **probably** working with an exponential model.

**Pattern examples**

| | |
|---|---|
| According to the US Center for Disease Control, the numbers of new cases of COVID-19 reported to them each day during September through December of 2020 were approximately:<br><br>Sep 2020 44,000 new cases per day<br>Oct 2020 75,000 new cases per day<br>Nov 2020 127,000 new cases per day<br>Dec 2020 216,000 new cases per day<br><br>Which of the following functions of the number of new cases $c(t)$ best models the pandemic in the US during this period of time (where $t$ is the number of months following September 2020)?<br><br>A) $c(t)=5,700t+44,000$<br>B) $c(t)=5,700+44,000t$<br>C) $c(t)=44,000(1.7)^t$<br>D) $c(t)=44,000(1.7)^{30t}$ | The first step is to decide whether the relationship between time and new cases is linear or exponential. Notice that time values are evenly spread (0, 44000), (1,75000), (2,127,000) and (3, 216,000). However, new cases are not growing by the same amount each month, therefore the relationship cannot be linear (this is a telltale signal of a pandemic).<br><br>We can eliminate (A) and (B) because they are linear.<br><br>The next step is to notice that the new cases are growing by the same rate.<br>$\frac{75}{44}=\frac{127}{75}=\frac{216}{127}=1.7$<br><br>SOLUTION: C |

**Comments on some official problems:**

1.1.24 You can see that the equation is a line because x increases by 5 when y increases by 30. That tells you the slope is 6. The translation down 5 units gives this question a twist.

Take any point and adjust it down 5, e.g., (18,130) becomes (18, 125). Use that point and the slope of 6 to get the equation for line K: y=6x+22. Set y=0 to find that x=-11/3.

1.2.17 The function increases by a rate, not by an amount, so it must be exponential. The problem already says that it is increasing.

1.2.24 Usually you would see two linear and two exponential models but these are all exponential models and they all have a base of 1.6. The general form of the exponential model is $f(x)=ab^x$. We are told that $f(1)=k=ab$. So now $kb^{-1}=a$ which means that the general form of the exponential becomes $f(x)=ab^x=kb^{x-1}$. The only answer choice with an exponent of $x-1$ is (C).

1.2.25 The hard part here is figuring out the base. We have a population increasing by 150% per year. So the base is $b=(1+r)=\left(1+\frac{150}{100}\right)=2.5$. To find the starting amount $a$ just solve $180=a(2.5)^1$.

2.2.17 8,000 is the starting amount in the starting year. You don't need this for the problem, but note that the base is .65 which means exponential decay at the rate of 35%.

3.1.23 The easy way to solve this is to recognize that the equation is $y=2.01x$, which is linear because $x$ is not raised to a power. Or you could make up (x,y) pairs (1, 2.01), (2, 4.02), (3, 6.03) etc. and notice that y is increasing by the same amount.

3.2.17 Notice that $p(1)$ must be same as $r(12)$ and then find a match.

3.2.22 Notice you are given the starting value $a=200$ and the base $b=4$. Then $y=ab^x=200(4)^x$.

4.2.23 We know that the starting amount is 5,470. After one year that amount is f(12)=5470(.64). So $b=.64=1-r$.

## 3.8 When given the coordinates of a point that lies on a curve plug those coordinates into the equation of the curve

**Why this is awkward:** Although this method is used in school, it still can be hard to spot. It is especially awkward when this method is used to solve for a variable. To make it more difficult, the data point can be used to solve for one variable in an expression that contains multiple other variables.

**How to spot the pattern:** Notice that you are being given the coordinates of a point. This information may be conveyed via:

- An (x,y) coordinate pair. This is the easiest to spot. For example, "the graph of the polynomial travels through the point (2,5)".
- A phrase. For example, "ten minutes after she left her home, Erin was cycling at the speed of 3 miles per hour (mph)." This is equivalent to being given the data point (10,3).
- A function value. For example, $g(4)=5$. This is equivalent to being given (4,5) as a data point.

**Pattern examples**

If $f(x)=2x^2+x+a$ and the graph of $f(x-a)$ passes through the point (2,20), then what is the value of $a$ ?

A) -1
B) 0
C) 1
D) 2

The first step is to find an expression for $f(x-a)$:

$f(x)=2x^2+x+a$

$f(x-a)=2(x-a)^2+(x-a)+a=2x^2-4ax+2a^2+x$

The second step is to use the point to solve for $a$,

at (2,20) $20=2(4)-4a(2)+2a^2+2$

$0=2a^2-8a-10=2(a-5)(a+1)$

$a=5$ or $a=-1$

SOLUTION: A

Calculator Tip:

Enter the equation for $f(x-a)$ replacing X with 2 and also replacing $a$. with X. We replace X with 2 because we are interested in testing for (2, 20). We replace $a$ with X because the calculator cannot solve for a variable other than X.

Press the *2nd-graph* key to see the table. We can backsolve values of $a$ by looking at the values of X in the table. The answer is (A) because that produces a Y value of 20.

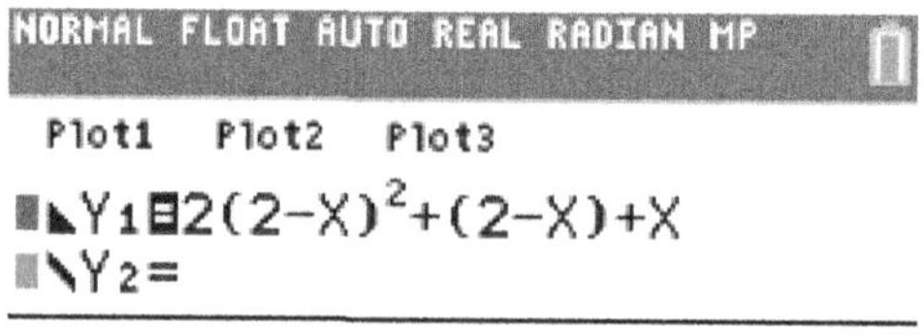


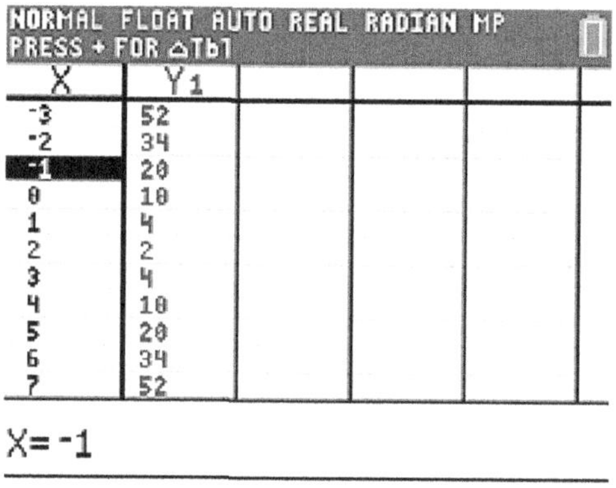

**Online Calculator Tip**

Enter the equation for $f(x-a)$ into the equation window. You will be prompted for a slider for $a$.

Add the slider. We will start backsolving the answer choices by setting the slider to the first answer choice, -1.

Notice that the curve travels through (2,20). Therefore (A) is correct.

You could check the next answer choice by moving the slider to zero.

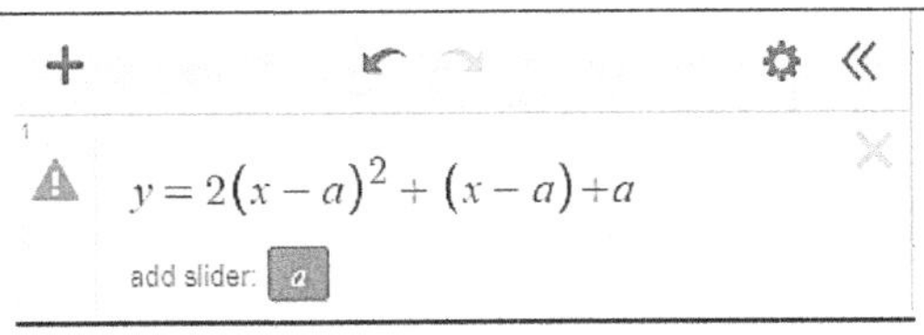


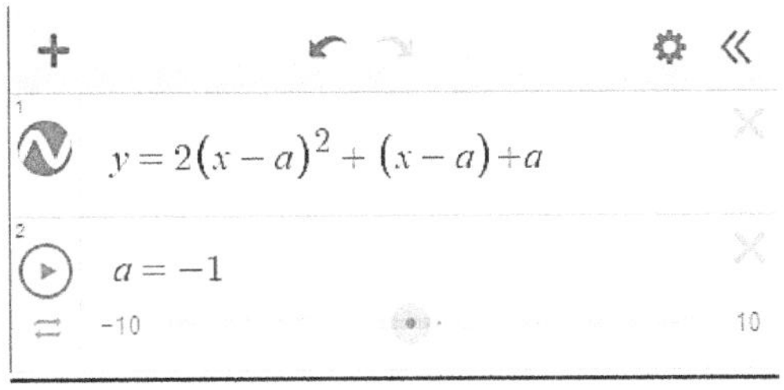


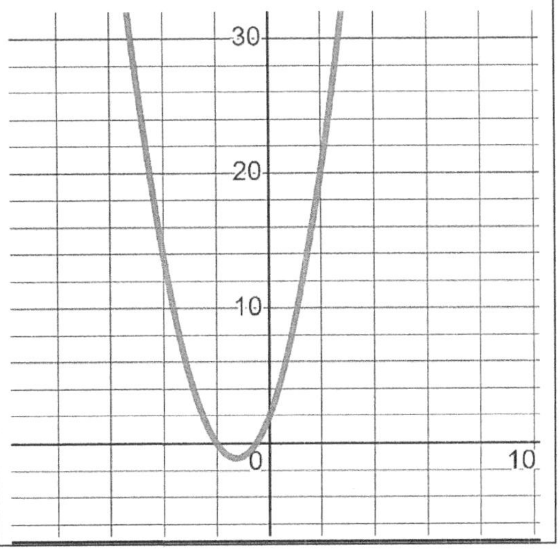

| | |
|---|---|
| If the graph of $2(2x-a)(x+b)$ passes through the point (2,6) then $a$ must be:<br><br>A) $-12$<br>B) $\frac{12b+71}{b+6}$<br>C) $\frac{4b+4}{b+2}$<br>D) $\frac{4b+5}{b+2}$ | The first step is to plug the point into the equation:<br>$6=2(4-a)(2+b)$<br>$3=(4-a)(2+b)=-2a-ab+4b+8$<br>$0=-a(2+b)+4b+5$<br><br>The second step is to solve for $a$:<br>$a(2+b)=4b+5$<br>$a=\frac{4b+5}{b+2}$<br><br>SOLUTION: D<br><br>**Solve by substitution and backsolving**<br><br>The problem places no restrictions on $b$ so let $b=0$. Then $2(2x-a)(x+b)$ becomes $2x(2x-a)$. Plug in the point (2,6) to get $6=4(4-a)$ or $5/2=a$. The last steps are to replace $b$ with zero in the answer choices until you find an answer choice with the value 5/2.<br><br>SOLUTION: D |

| Online Calculator Tip | Enter $2(2x-a)(x+b)$ into the equation window and choose to add sliders for $a$ and $b$.<br><br>The problem places no restrictions on $b$ so set that slider to zero.<br><br>Move the slider for $a$ around until you see the graph go through (2,6). | 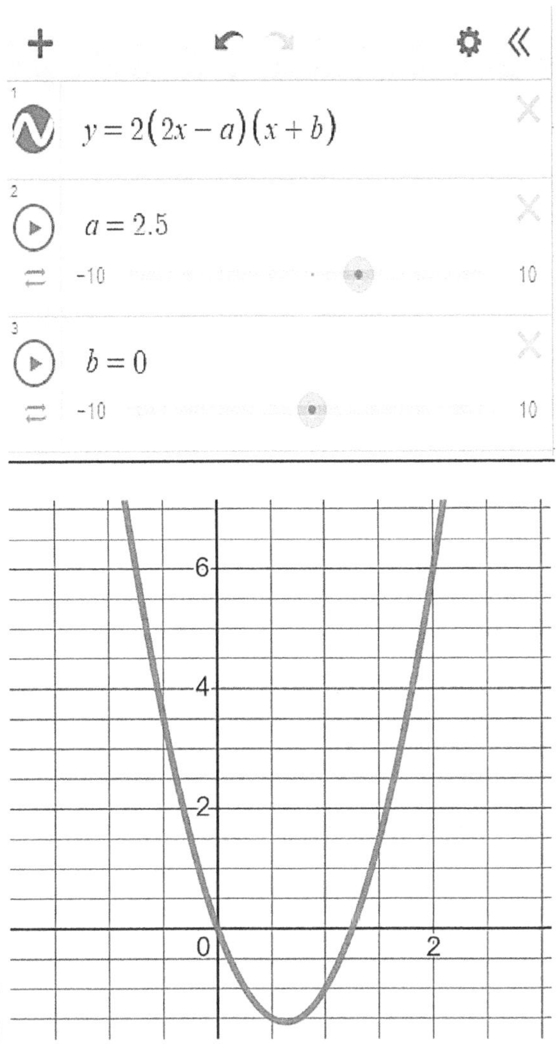 |
|---|---|---|



**Comments on some official problems:**

2.1.19 Using (a,0) we get $h(a)=4a+28=0$. Using (0,b) we get $h(0)=0+28=b$

2.2.24 Notice that you are given the point (0,y). Plug that point into the answer choices and find the answer choice that gives only one solution.

4.1.27 Plug $\left(0,\dfrac{-99}{7}\right)$ into y to find f(0), then use that to find $b=13/7$. Then solve $ab=\dfrac{65}{7}$ to find $a$.

# 4. Less remembered math facts

There are some math facts that I feel compelled to mention because these appear on the digital SAT® even though they may have received little emphasis or were not covered at all in school. In particular, statistics/probability questions have become more popular on standardized tests. That topic gets only a little attention outside a statistics class and it is something that students should not ignore. You might want to do some content review in probability and statistics, which will require another resource with lots of practice problems.

## 4.1 Percentage change

You may think that percentage changes are way too simple to worry about, but the people responsible for the digital SAT® can make this difficult. Percentage changes often appear in word problems, but not always. Percentage change can be tested in very creative ways.

Sometimes it is useful to use $\left(\frac{\text{new-old}}{\text{old}}\right)100$ to calculate a percentage change. If the value is positive, you have a percent increase, and if the value is negative, you have a percentage decrease.

These formulas will always work and are generally preferred. To increase a quantity $x$ by $p\,\%$ use $x\left(1+\frac{p}{100}\right)$; to decrease a quantity $x$ by $p\,\%$ use $x\left(1-\frac{p}{100}\right)$.

**Examples**

| In basketball, Mary shot 100 baskets so far this season. At this time last season she had shot 120 baskets. What is the percentage change that Mary had in baskets year over year?<br><br>A) 17% increase<br>B) 17% decrease<br>C) 20% increase<br>D) 20% decrease | You could use:<br><br>$\left(\frac{\text{new-old}}{\text{old}}\right)100=\left(\frac{100-120}{120}\right)100=-16.7$<br><br>Or you could use:<br>$100=\left(1-\frac{p}{100}\right)120$<br><br>SOLUTION: B |
|---|---|

| | |
|---|---|
| A store clerk was instructed to mark everything up by 15%. Instead, he marked everything down by 15%. If a pair of shoes was originally marked \$80, what is the difference (in dollars) between the correct price and the incorrect price marked by the clerk?<br><br>A) \$8<br>B) \$12<br>C) \$24<br>D) \$48 | The price should have been marked $80(1+.15)=92$ but instead it was marked $80(1-.15)=68$. The difference is $92-68=24$.<br><br>SOLUTION: C |

| | |
|---|---|
| If $y$ is a 30% increase over $x$ and $x$ is a 20% decrease below $z$, then<br>A) $y=(1.3)(0.8)z$<br>B) $y=(1.8)(0.7)z$<br>C) $y=(1.3)(1.2)z$<br>D) $y=(0.7)(0.8)z$ | $y$ is a 30% increase over $x$ is: $y=(1+.3)x$<br>$x$ is a 20% decrease over $z$ is: $x=(1-.2)z$<br>Together, this gives $y=(1.3)(0.8)z$<br><br>SOLUTION: A |

**Comments on some official problems:**

2.1.27 Use the percentage change formula to set up a system of equations

$a=\left(1+\frac{110}{100}\right)b$ and $b=\left(1-\frac{90}{100}\right)47$. Or you can build a system using the general formulas:

$a=\left(1+\frac{110}{100}\right)b$ and $b=\left(1-\frac{90}{100}\right)47=4.7$.

3.1.21 80% off the regular price is $11.70\left(1-\frac{80}{100}\right)=2.34=\text{sale price}$. Next

$2.34=\left(1+\frac{30}{100}\right)\text{cost}$

4.2.27 Solve for p: $210=\left(1+\frac{p}{100}\right)30$

## 4.2 Descriptive statistics and their graphs

Here, we review statistics that can be used to describe a data set: mean, median, mode, range, and standard deviation. The mean, median and mode are measures of central tendency, whereas the range and standard deviation are measures of spread.

The mean is the average, $mean = \frac{sum}{count}$ .

The median is the "middle number." If the count is odd, the median will actually be the middle number in the data set. If the count is even the median is the average of the two middle numbers.

The mode is the number that appears most frequently in the data set. There can be no mode (where all numbers appear with the same frequency). There can be one mode (one number appears most frequently), or there can be two or more modes (two or more numbers appear with the same frequency and that frequency is more than any of the other frequencies in the data set).

The definition of the range is that it is the largest number in the data set minus the smallest number in the data set, or

$$\text{range} = \text{maximum value - minimum value}$$

Think of the standard deviation as a measure of how tightly the numbers are grouped around the mean. The smaller the standard deviation is, the more tightly the numbers are grouped around the mean.

**Examples**

| | |
|---|---|
| If a data set consists of the values<br>-2, 0, -6, 7, 9, -3<br>What is the value of the mean subtracted from the median?<br><br>A) -11/6<br>B) -1/6<br>C) 1/6<br>D) 11/6 | First, order the data set:<br>-6, -3, -2, 0, 7, 9<br><br>The mean is 5/6 and the median is -1.<br><br>$-1-(5/6)=-11/6$<br><br>SOLUTION: A<br><br>Note that the range is $9--6=15$ and there is no mode. |

| | | |
|---|---|---|
| Calculator Tip: | Enter stat-edit to enter the data set into the calculator.<br><br>Press stat-calc-1-Var Stats to get the descriptive statistics. The mean is shown on the first line. You can find the median by scrolling down. Although not needed, $Sx$ and $\sigma x$ are two forms of the standard deviation. | 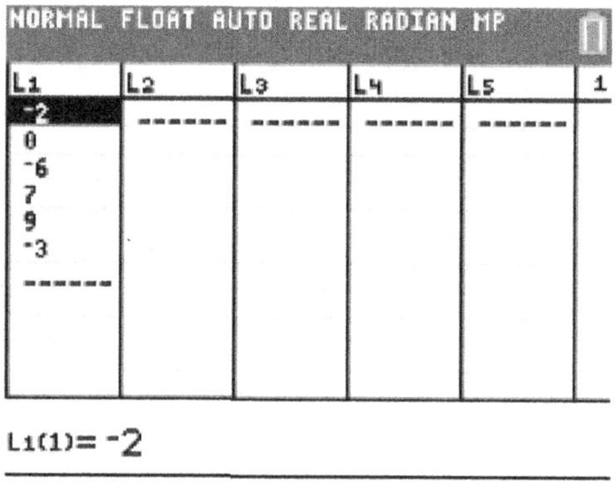<br>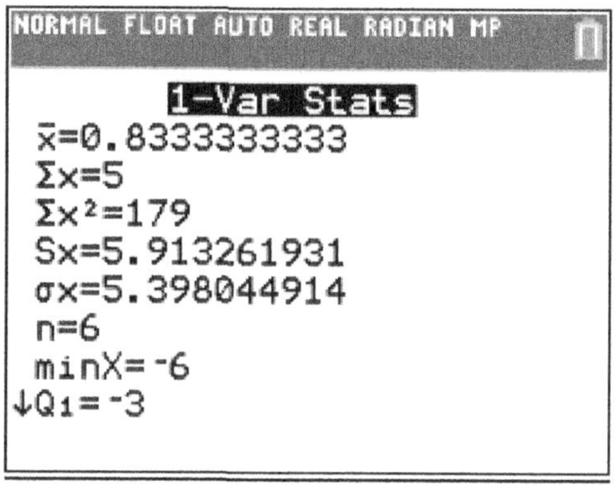 |

| | | |
|---|---|---|
| Online Calculator Tip | Click on the Plus button on the top left to add an object and choose table. The table will have two columns named $x_1$ and $y_1$. Enter the data values into the first column.<br><br>Click on the Plus button and add an expression. Click on the keyboard icon and then click on functions. Choose mean and fill it with x1.<br><br>Repeat this process for the median. | 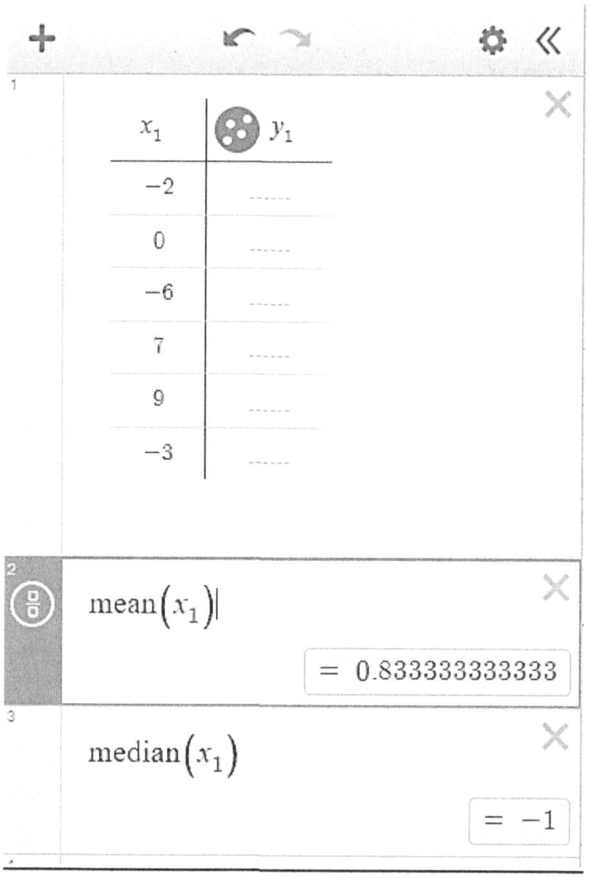 |

Suppose you are responsible for purchasing screens for Apple's iPhone. There are two suppliers whose contracts are up for renewal. Their screen production is shown below.

The goal is to produce a screen with a width of 0.1 mm. Results from the two suppliers have the same mean, the same median, and the same mode of 0.1 mm. They also have the same range of about .21 mm.

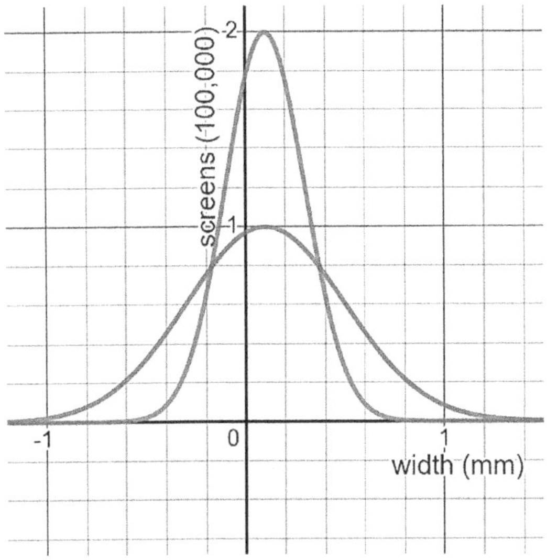


Assuming that costs are the same, which supplier contract should you renew, the one with the taller graph or the one with the shorter graph?

A) the taller graph

B) the shorter graph

C) cannot be determined because the two suppliers are statistically equivalent

What differs statistically is their standard deviation. The taller graph has less spread about the mean than the shorter graph. Therefore, the standard deviation is smaller for the taller graph than the shorter graph. The taller graph is the better supplier because more of its screens are closer to the 0.1 mm target width.

SOLUTION: A

Some questions may require you to read a dot plot, which is easy to do. For example Data Set A consists of the values {-5, -5, -5, -5, -3, -1, 0, 4, 4, 4, 5, 5}.

The dot plots below illustrate data values collected from Study A and Study B.

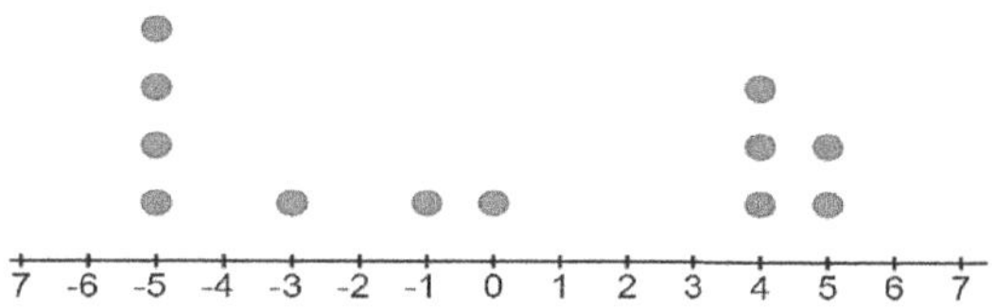


Data Set A

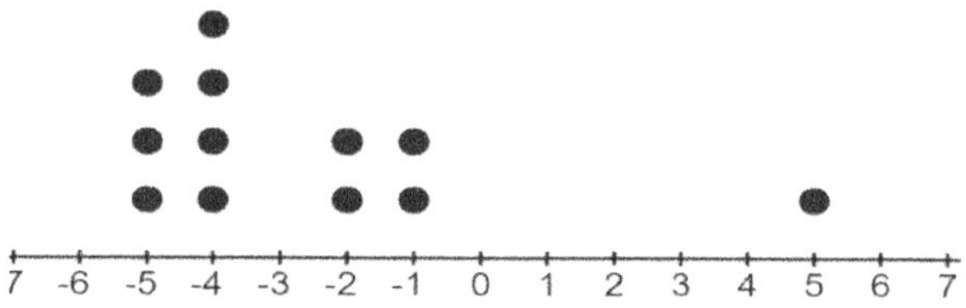


Data Set B

Which of the following statements must be true?

I. The median of Data Set A is equal to the median of Data Set B.

II. The standard deviation of Data Set A is equal to the standard deviation of Data Set B.

A) I only
B) II only
C) I and II
D) Neither I nor II

Make certain that you know how to read a dot plot. Data Set A consists of 12 data points and their values are -5, -5, -5, -5, -3, -1, 0, 4, 4, 4, 5, 5.

The median for Data Set A looks to be approximately 0 whereas the median for Data Set B looks to be approximately -4. Therefore, the medians are not equal.

Data Set A has two clusters of data points and both clusters are far away from the mean, whereas Data Set B has one point far away from a cluster that contains all the other points in close proximity to the mean. Therefore, Data Set A has a larger standard deviation than Data Set B because Data Set A has fewer points grouped in the neighborhood of the mean.

SOLUTION: D

Some questions may require you to read a box plot (think of a cat with a box for a mouth and whiskers). As you can see below, the first vertical stripe is the value of the smallest value in the data set. The second vertical stripe is the first quartile of the data set, meaning that 25% of the data points lie below the second stripe. The third vertical stripe is the median of the data set. The fourth vertical stripe is the third quartile of the data set, meaning that 75% of the data lie below the fourth vertical stripe. The fifth vertical stripe is the largest value in the data set.

In an actual box plot, the minimum, quartiles and maximum will not be labeled as such. **You must memorize what each of the vertical stripes represent.**

In Data Set B, the minimum is -5, the first quartile is approximately -4, the median is approximately -3.5, the third quartile is approximately -1 and the maximum is 5. In Data Set A, the minimum value and the value of the first quartile are the same.

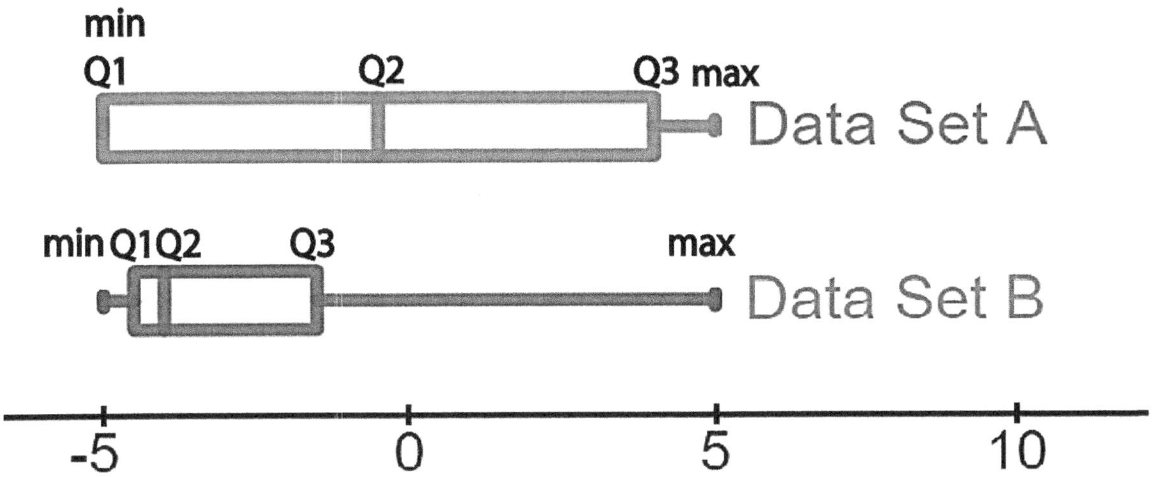


| Which of the statements below cannot be determined?<br><br>A) The median of data set A is larger than the median of data set B.<br>B) The mode of data set A is larger than the mode of data set B.<br>C) The standard deviation of data set A is larger than the standard deviation of data set B.<br>D) The range of data set A is larger than the range of data set B. | A) The median Q2 for data set A is clearly larger than the median Q2 for data set B.<br><br>B) Box plots do not provide any information about the mode.<br><br>C) It is clear the that data points are more tightly clustered around the mean in data set B than they are in data set A.<br><br>D) The ranges of the data sets are the same<br><br>SOLUTION: B |
|---|---|

**Comments on some official problems:**

3.2.20 In the numerator, use the sum for A and the sum for B:

$$\text{mean} = \frac{\text{sum}}{\text{count}} = \frac{75(32)+50(62)}{75+50}$$

3.2.24 Use the strikeout method to find that 13 is the median for both data sets. Notice that that the data points cluster more tightly around the median for data set A versus data set B. Therefore the standard deviation for data set A is less than the standard deviation for data set B.

4.2.24 You could write a list of the original 15 values and another list of the 15 values when 56 is added. However it should be clear that the new median is just 56 added to the old median. It should also be clear that the range does not change because
(new max)-(new min)=(old max+56)-(old min+56)

## 4.3 Calculating descriptive statistics from a frequency table or histogram

It can be a little puzzling when asked to calculate statistics based on a frequency table or histogram, rather than using a dot plot or a list of data points. We start by taking a look at a frequency table. The method for a histogram is the same.

**Examples**

In math class the teacher asked every student to find out how many TV sets are in their household. The results from the 39 students are shown below.

| Households | TV Sets |
|---|---|
| 2 | 0 |
| 3 | 1 |
| 8 | 2 |
| 12 | 3 |
| 9 | 4 |
| 5 | 5 |

What is the result when the median number of TVs per household is subtracted from the mode number of TVs per household?

A) 0
B) 3
C) 4
D) 12

First you must understand how to read the table. The underlying data set looks like 0,0,1,1,1,2,2,2,2,2,2,2,2, etc.

Finding the mode is easy, that is the most frequent value, which is 3 TV sets. Finding the median is a bit more tricky. We use a different sort of strike out method:

Use the frequencies:

**2, 3, 8, 12, 9, 5**

Then start by striking out 5 from the bottom and 5 from the top.

**~~2~~, ~~3~~, 8, 12, 9, ~~5~~**

The next striking out is a little tricky, striking out 8 from the top and 8 from the bottom:

**~~2~~, ~~3~~, ~~8~~, 12, ~~9~~, ~~5~~**
**~~2~~, ~~3~~, ~~8~~, 12, 1, ~~5~~**

The last striking out is

**~~2~~, ~~3~~, ~~8~~, 12, 1, ~~5~~**
**~~2~~, ~~3~~, ~~8~~, 11, ~~1~~, ~~5~~**

This shows that after all the striking out, we have eleven data values of 3 TV sets. Therefore the median is 3.

The mode minus the median is $3-3=0$.

SOLUTION: A

Note that the mean is

$$\frac{2(0)+3(1)+8(2)+12(3)+9(4)+5(5)}{2+3+8+12+9+5}=\frac{116}{39}$$

The range is $5-0=5$.

**Calculator Tip:**

Enter stat-edit to enter the data set into the calculator.

Press stat-calc-1-Var Stats to get the descriptive statistics. Before hitting calculate make sure that the data value list is $L_2$ and the frequency list is $L_1$ (choose these using 2nd-1 and 2nd-2)

Now highlight calculate and press enter. Scroll down to find the median (Med). The calculator does not display the mode or range.

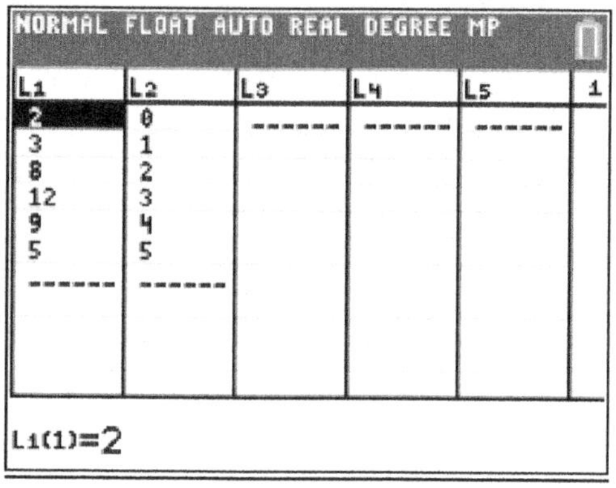


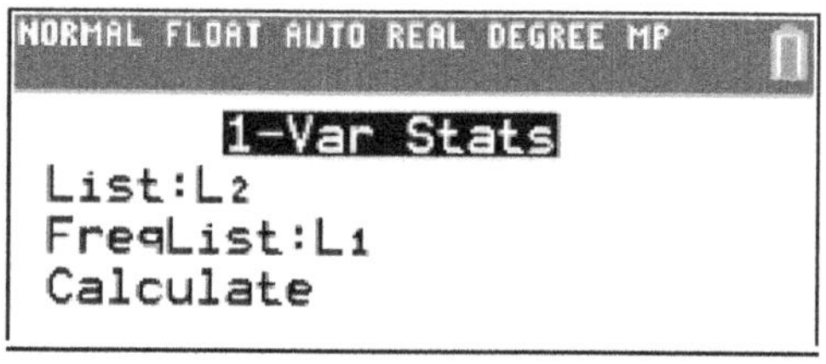


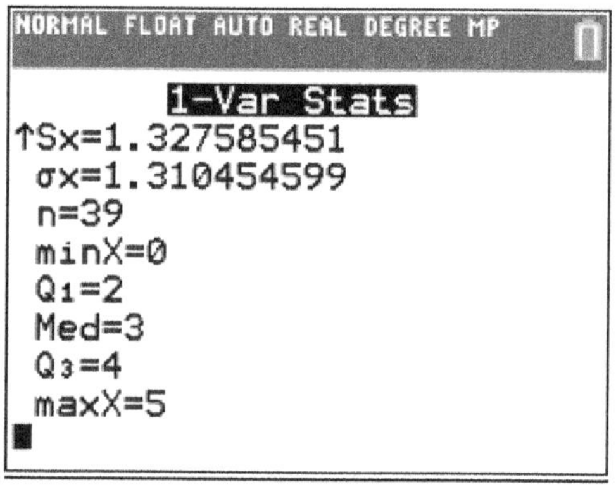

**Online Calculator Tip**

Click on the Plus button on the top left to add an object and choose table. The table will have two columns named $x_1$ and $y_1$. Enter the data values given in the table.

Click on the Plus button and add an expression. Shown below is how you could calculate the mean of the data set by using the total function, which can be found in the stats section of the function keypad.

Click on the Plus button and add an expression. Shown below is how you could calculate the range of the data set using the max and min functions that can be found in the stats section the keypad.

To find the median, use the strikeout method discussed earlier or note that there are 39 households. 39/2 is about 20, so starting with the top of the table cross out 2 and then 3 and then 8. That gives 13 TVs accounted for with 7 not accounted for. Those 7 must lie in the next row because there are 12 TVs in that row. Therefore, the median must be 3.

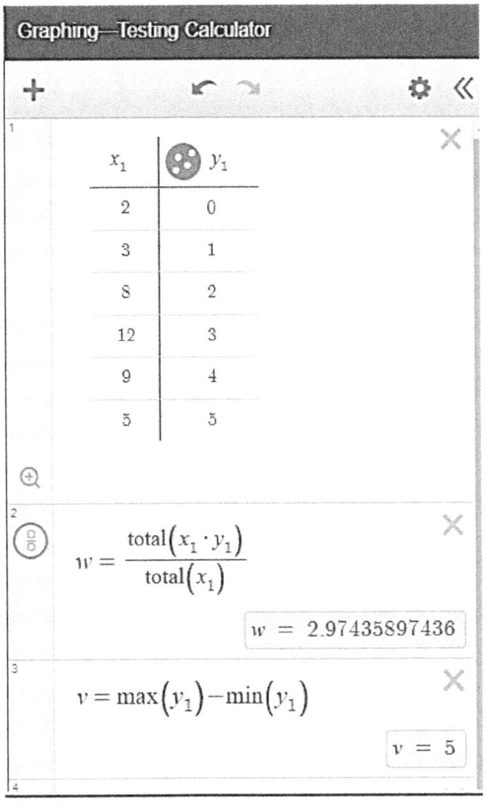

In math class the teacher asked every student how many TV sets are in their household. The results were as follows

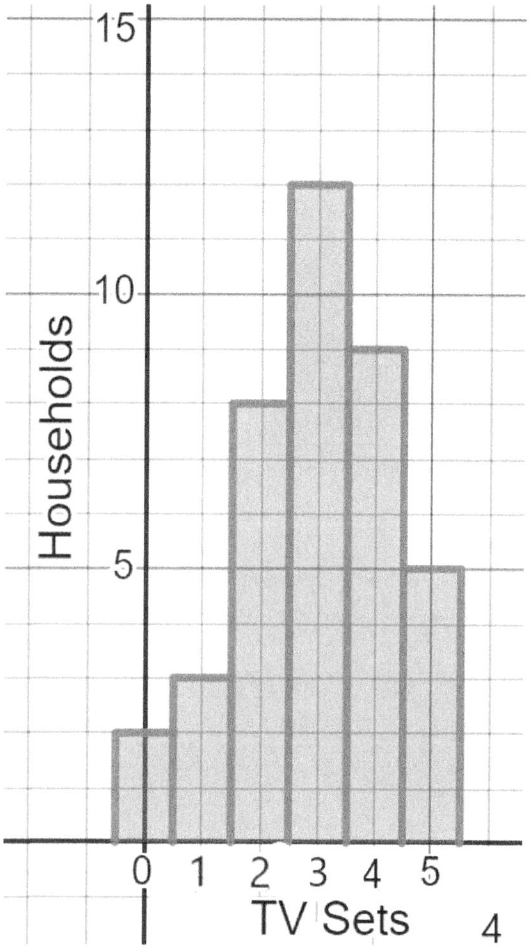


As you can see, some households had no TV sets where others had five TV sets.

What is the relationship between the mean and the median numbers of TV sets per household?

A) The mean is greater than the mean

B) The mean is equal to the median

C) The mean is less than the median

D) It cannot be determined

First you must understand how to read the table. The underlying data set looks like 0,0,1,1,1,1,2,2,2,2,2,2,2,2, etc.

Finding the median is a bit tricky. We use a different sort of strike out method:

Use the household frequencies:

**2, 3, 8, 12, 9, 5**

Then start striking out 5 from the bottom and 5 from the top.

**~~2~~, ~~3~~, 8, 12, 9, ~~5~~**

The next striking out is a little tricky, striking out 8 from the top and 8 from the bottom:

**~~2~~, ~~3~~, ~~8~~, 12, ~~9~~, ~~5~~**
**~~2~~, ~~3~~, ~~8~~, 12, 1, ~~5~~**

The last striking out is

**~~2~~, ~~3~~, ~~8~~, 12, 1, ~~5~~**
**~~2~~, ~~3~~, ~~8~~, 11, ~~1~~, ~~5~~**

This shows that after all the striking out, we have eleven data values of 3 TV sets. Therefore the median is 3.

The mean is

$$\frac{2(0)+3(1)+8(2)+12(3)+9(4)+5(5)}{2+3+8+12+9+5}=\frac{116}{39}=2.97$$

SOLUTION: C

Note that the range is $5-0=5$ and the mode is 3 (very easy to spot on a histogram).

**Calculator Tip:**

Enter stat-edit to enter the data set into the calculator.

Press stat-calc-1-Var Stats to get the descriptive statistics. Before hitting calculate make sure that the data value list is $L_2$ and the frequency list is $L_1$ (choose these using 2nd-1 and 2nd-2)

Next highlight calculate and press enter. The mean ($\bar{x}$) is on the first line. Scroll down to find the median (Med). The calculator does not display the mode or range.

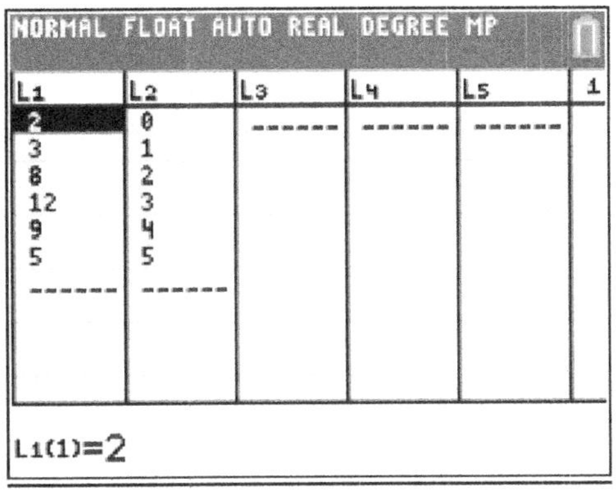


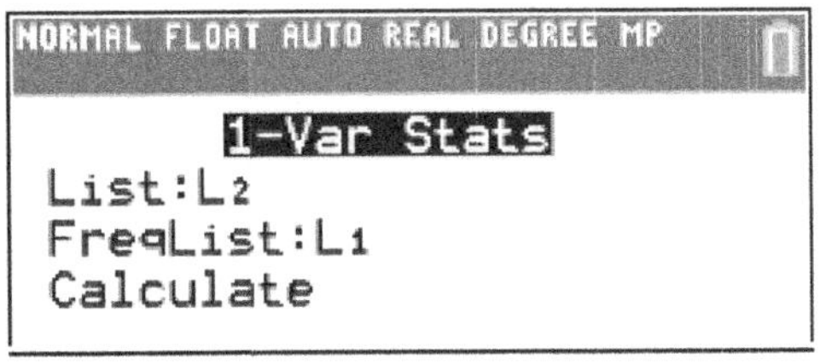


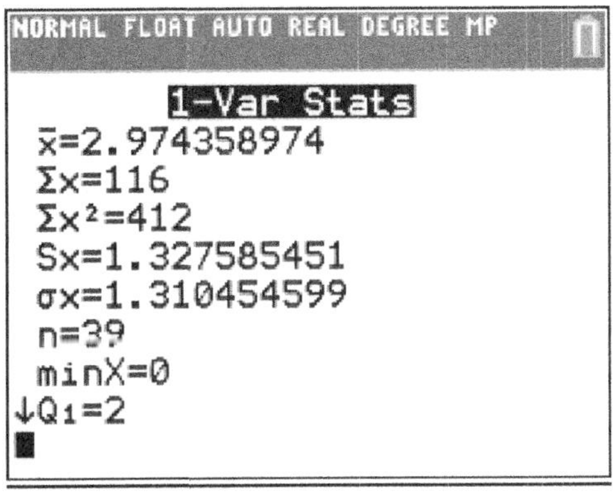

**Online Calculator Tip**

Click on the Plus button on the top left to add an object and choose table. Enter the data values into the columns.

Click on the Plus button to add an expression. Shown below is how you could use the total function to calculate the mean of the data set. The total function is in the stats section of the keypad.

Click on the Plus button and add an expression. Shown below is how you could calculate the range of the data set using the max and min functions which are in the stats section of the keypad.

To find the median, use the strikeout method discussed earlier or note that there are 39 households. 39/2 is about 20, so starting at the top of the table cross out 2 then 3 and then 8. That gives 13 TVs accounted for. The other 7 must lie in the next row because there are 12 TVs in that row. Therefore, the median must be 3.

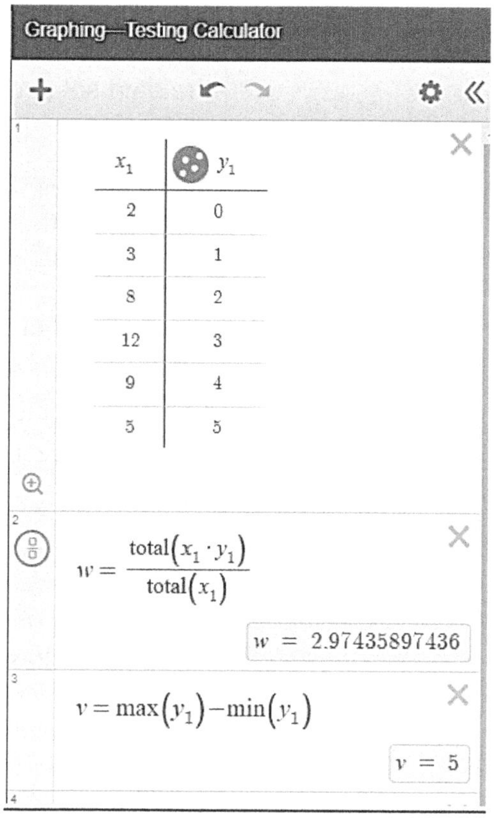


**Comments on some official problems:**

2.2.26 To find the smallest difference, use min A – max B. To find the smallest mean for A, use the left endpoints to find the mean:

$$\frac{3(20)+4(30)+7(40)+9(50)}{23}=\frac{910}{23}$$

To find the largest mean for B, use the right endpoints to find the mean:

$$\frac{3(19)+4(29)+7(39)+9(49)}{23}=\frac{867}{23}$$

## 4.4 Joint and conditional probabilities

Of the various probability questions that may arise, joint and conditional probabilities are the most frequent. A joint probability is the probability of two outcomes occurring at the same time. A conditional probability is the probability of one outcome occurring after another outcome has occurred. The example below will help with these concepts.

**Example**

Suppose there are 150 seniors about to graduate from Woods High School and that every graduating senior was asked whether they plan to attend college. Their responses are tabulated below.

| | Do you plan to attend college? | | |
|---|---|---|---|
| | Yes | No | Total |
| Black | 25 | 25 | 50 |
| White | 70 | 30 | 100 |
| Total | 95 | 55 | 150 |

**Joint Probability Question and Answer**

If a graduating senior at Woods High School is chosen at random, what is the probability that the student is black and planning to attend college?

A) 0.17
B) 0.47
C) 0.50
D) 0.70

We are asked the probability of a graduating senior is black and planning to attend college. Therefore, the denominator is the total number of graduating seniors. 25/150

SOLUTION: A

The question could also have been worded "What is the probability that a graduating senior is black and planning to attend college?".

Notice that the key word is "and".

**Conditional Probability Question and Answer**

What is the probability that a black graduating senior at Woods High School would plan to attend college?

A) 0.17
B) 0.47
C) 0.50
D) 0.70

We are asked the probability that a black graduating senior plans to attend college, therefore the denominator is 50. 25/50

SOLUTION: C

The question could have been worded "What is the probability of a graduating senior planning to attend college given that s/he is black?".

Notice that the key word is "given".

**Comments on some official problems:**

4.2.13 This is a joint probability of 3/10.

## 4.5 Census vs. a sample, and margin of error

Suppose you live in a town of 10,000 people and you want to know what percentage of people would support making repairs to the local bridge. If you wanted an exact value for the percent in favor, you would make sure that you asked every person. That is called a census.

A census is expensive and time consuming, so instead you decide to ask just 500 people, selected randomly. That is called a sample. With a sample you will save time and money but you will also introduce some error in your estimate of the percentage of all the residents in favor of funding repairs to the bridge. You will know the exact percentage of your sample in favor, but because you did not ask everyone, there will be a margin of error in your estimate.

You do not need to know how to calculate the margin of error. All you need to know is that you can be confident that the percentage of the entire population of residents in your town in favor of repairing the bridge will be between the estimated percentage found in your sample plus or minus the margin of error. For example, if your sample found 60% in favor with a margin of error of 5%, you would be confident in estimating that the percentage of all residents in favor ranged between 55% and 65%. Also, although intuitive, you need to know that the margin of error can be reduced by increasing the size of your sample.

**Examples**

<table>
<tr>
<td>Izzy is considering whether to run for president of her student council. To help her decide, she asks a random sample of students whether they would vote for her. In the sample, 70% said they would vote for her and the margin of error is 10%. Based on these results what can be concluded about Izzy's prospects for election?<br><br>A) If she runs, Izzy will receive exactly 70% of the votes.<br>B) If she runs, Izzy will likely receive at least 80% of the votes.<br>C) If she runs, Izzy will likely receive between 60% and 80% of the votes.<br>D) If she runs, Izzy will likely receive fewer than 60% of the votes.</td>
<td>The estimated percentage of votes Izzy will receive is likely to be between the sample percentage plus or minus one margin of error, in this case 70% plus or minus 10%.<br><br>SOLUTION: C</td>
</tr>
</table>

| | |
|---|---|
| The mayor of AnyTown is trying to decide where to place a new cell phone tower. He decided to choose a random sample from the population of 5,000 people, and ask their preference. The results were as follows:<br><br>Northern part of town 150<br>Central part of town 120<br>Southern part town 80<br><br>If the entire population had voted on the decision, how many more residents would be expected to vote in favor of the northern location versus the number expected to vote for the southern location?<br><br>A) 30<br>B) 70<br>C) 429<br>D) 1000 | The sample consisted of 350 people. Of those included in the sample, a difference of 150-80 people favored the northern location versus the southern location. The expected number for the population would be $\left(\frac{150-80}{350}\right)5000=1000$<br><br>SOLUTION: D |

**Comments on some official problems:**

3.1.26 The estimate of votes for Angel Cruz would be $\left(\frac{483}{483+320}\right)6,424$.

4.2.10 The proportion would likely fall in the interval of $0.49 \pm .04 = 0.45$ through $0.53$.

## 4.6 Equation of a circle

The equation of a circle is $(x-h)^2+(y-k)^2=r^2$, where the center of the circle is at $(h,k)$ and the radius is $r$. These questions can have several variations.

### Examples

If a circle is tangent to the x-axis at x=5 and is tangent to the y-axis at y=-5, what is the equation of the circle?

A) $(x+5)^2+(y+5)^2=25$

B) $(x+5)^2+(y-5)^2=25$

C) $(x-5)^2+(y+5)^2=25$

D) $(x-5)^2+(y-5)^2=25$

The answer choices differ with respect to the center. Because of the positions of the intercepts, the center of the circle must be at (5, -5). Therefore, the equation must involve $x-5$ and $y+5$.

SOLUTION: C

Online Calculator Tip

Backsolve the answer choices using the online calculator to find the graph with tangent points at (5, 0) and (0, -5).

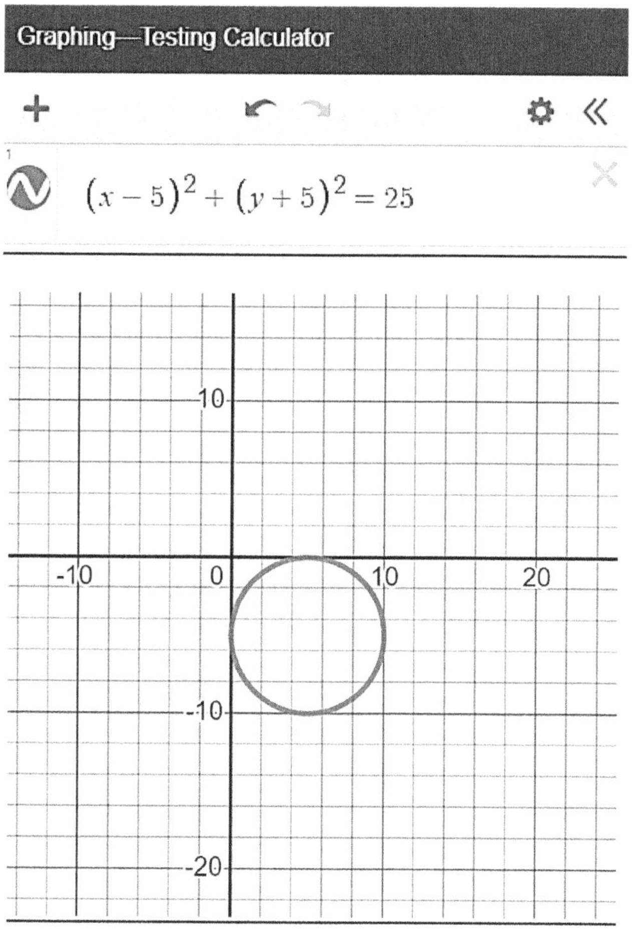

What are the coordinates of the center of a circle whose equation is

$x^2 + 6x + y^2 - 2y = 6$ ?

A) (6, -2)
B) (-6, 2)
C) (3, -1)
D) (-3, 1)

You must complete the square, using the formula $\left(\frac{b}{2}\right)^2$ to create perfect squares.

$$x^2 + 6x + y^2 - 2y = 6$$
$$(x^2 + 6x + 9) + (y^2 - 2y + 1) = 6 + 9 + 1$$
$$(x + 3)^2 + (y - 1)^2 = 16$$

SOLUTION: D

Online Calculator Tip

Enter the equation of the circle, and notice that the center is at (-3, 1).

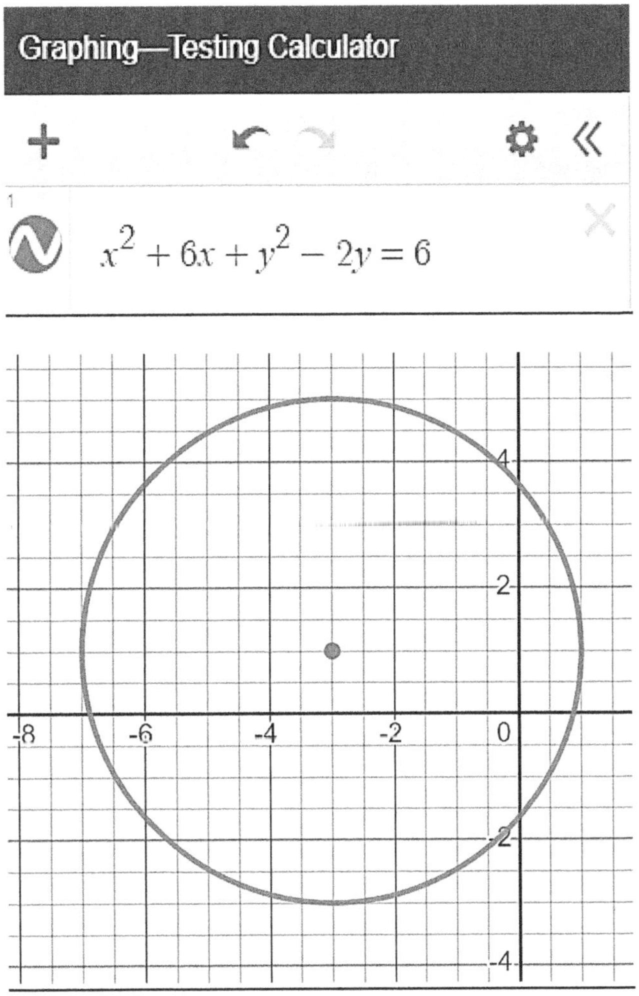

| What is the center of a circle whose radius is $2\sqrt{2}$ and travels through the point (3, -1) ?<br><br>A) (3, -1)<br>B) (1, -3)<br>C) (-3, -1)<br>D) (-1, -3) | Start by plugging in the point (3, -1) and plug in the radius squared:<br><br>$(3-h)^2+(-1-k)^2=\left(2\sqrt{2}\right)^2=8$<br>Backsolve the various answer choices for $(h,k)$ until you find one that works:<br><br>$(3-1)^2+(-1--3)^2=2^2+2^2=8$<br><br>SOLUTION: B |
|---|---|

| Online Calculator Tip | Start by plugging in the point (3,-1) and the radius squared into the equation of the circle:<br><br>$(3-h)^2+(-1-k)^2=8$<br>Unfortunately Desmos will only graph what is called an "implicit equation" using the variables $x$ and $y$.<br><br>So enter this equation instead, replacing $h$ with $x$ and $k$ with $y$.<br>$(3-x)^2+(-1-y)^2=8$<br><br>Notice the center is at (3, -1). | 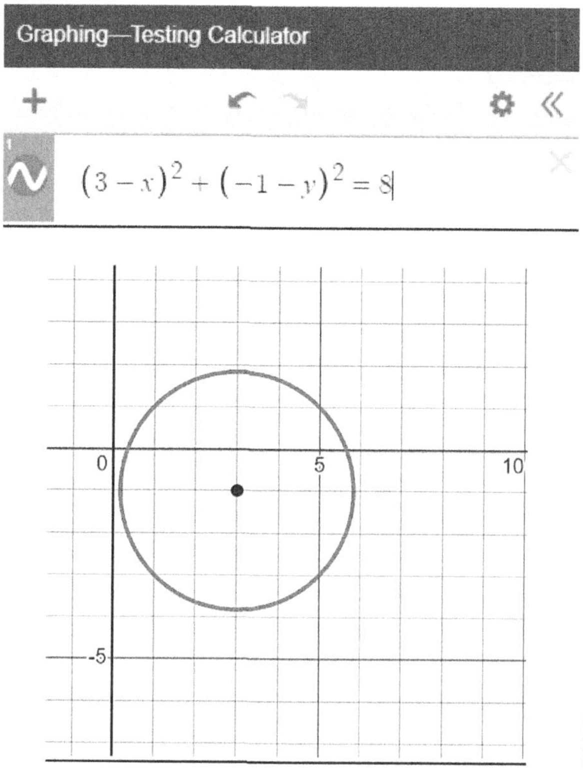<br> |
|---|---|---|

**Comments on some official problems:**

1.1.21 Find that the length of the diameter is 14-4=10.

1.2.23 Notice that there are answer choices for the radius and diameter.

2.2.24 Along the y-axis, x equals zero. Set x=0 and backsolve the answer choices. Find the answer choice that has only one solution.

3.1.27 Complete the square to find that $r^2=100$.

4.2.25 Move the original center point (0,1) down 2 units to (0, -1).

## 4.7 Circumference vs. area of a circle

Everyone has seen the formulas for the area and circumference of a circle, but so many students (at all levels) confuse them. Therefore, it is wise to do a quick refresh.

**Example**

<table>
<tr>
<td>A circle is inscribed in a square. If the side of the square measures 10 units, what is the area of the circle?<br><br>A) $5\pi$<br>B) $10\pi$<br>C) $25\pi$<br>D) $100$</td>
<td>This is a Geometry word problem and all of my students are repeatedly told that when they see a Geometry word problem they should draw a sketch.<br>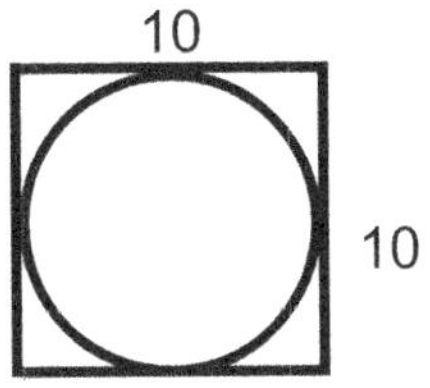<br><br>It should be apparent that the length (or width) of the square is the same as the diameter of the circle. Therefore, the radius of the circle is 5 and the area of the circle is $\pi r^2 = 25\pi$ .<br><br>SOLUTION: (C)<br><br>Answer (A) would trip up students who forget to square the radius. Answer (B) would trip up students who calculated the circumference of the circle. Answer (D) would trip up students who solved for the area of the square instead of the area of the circle.</td>
</tr>
</table>

**Comments on some official problems:**

1.1.19 There is an answer choice for the ratio of the circumferences (which is 43) as opposed to the ratio of the areas.

## 4.8 Complementary angles

In Geometry you were taught that if two angles are complementary, then their measures add up to 90 degrees. However, you might not remember that complementary angles have special properties. Rightly or wrongly, this topic can appear on the digital SAT©. The facts you need to remember are if $\angle A$ and $\angle B$ are complementary then:

$$\sin A = \cos B$$
$$\cos A = \sin B$$
$$\tan A = \cot B$$

It helps to know that complementary angle problems are usually rated hard and so tend to appear toward the end of the modules. These questions are actually easy to solve if you can remember (or derive) the identities above. Some ways complementary angles can occur are:

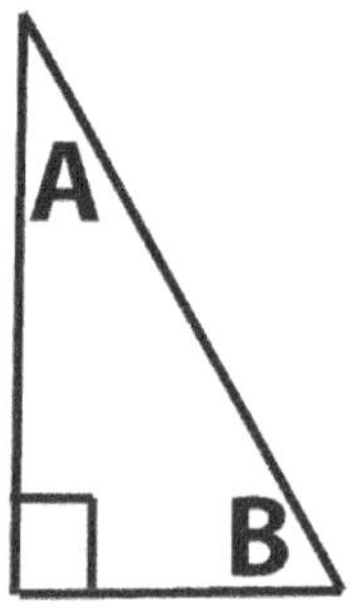


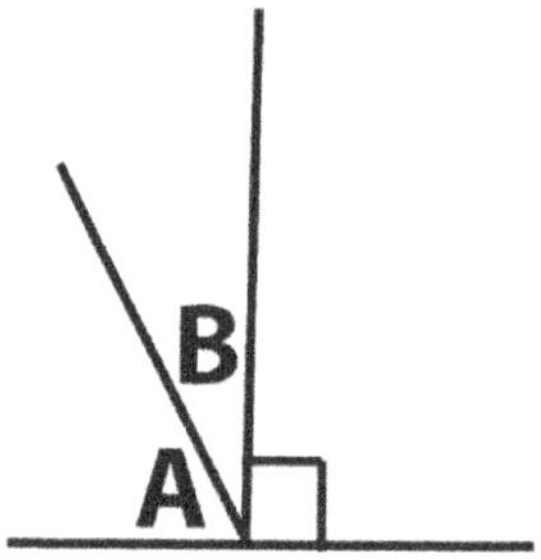


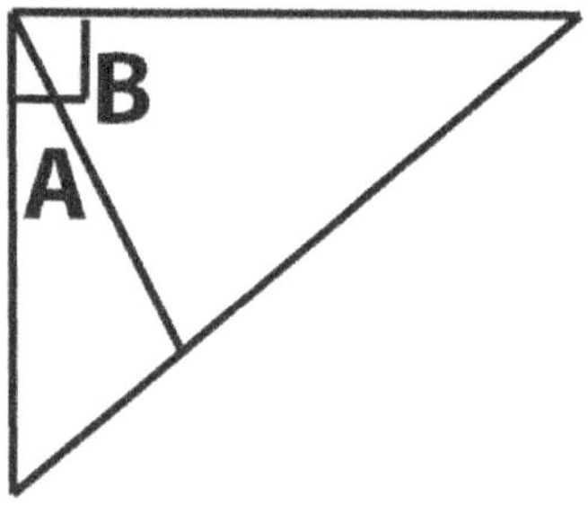


**Example**

| In triangle ABC,.angle B is a right angle. If $\sin A = 6/7$ then the $\cos C$ is<br><br>A) 6/7<br>B) 7/6<br>C) $\sqrt{13}/7$<br>D) $6/\sqrt{13}$ | If you recognize the complementary angles then you know that $\cos C = \sin A = 6/7$. However, a quick sketch of the right triangle would reveal the same thing.<br><br>SOLUTION: A |
|---|---|

**Comments on some official problems:** There are no instances of this in the linear pencil and paper practice tests, but there are some in the adaptive Bluebook™ practice tests.

## 4.9 Similar triangles

Most students remember spending loads of time proving that triangles are congruent, however many forget about similar triangles. The digital SAT© has some questions on congruent shapes but many more on similar shapes.

There also can be some confusion about the congruence symbol $\cong$ versus the similar symbol $\sim$ . If you see the statement $\triangle ABC \sim \triangle DEF$ it means that corresponding angles are congruent and corresponding sides are proportional. In this particular case, $\angle A \cong \angle D$ $\angle B \cong \angle E$ and $\angle C \cong \angle F$ and also $\frac{AB}{DE} = \frac{AC}{DF} = \frac{BC}{EF}$. To prove that triangles are similar, you can show that corresponding angles are congruent (AA), or corresponding sides are proportional (SSS). Similarity can also be proved by SAS, where the enclosed corresponding angles are congruent and the surrounding corresponding sides are proportional.

**Example**

<table>
<tr>
<td>

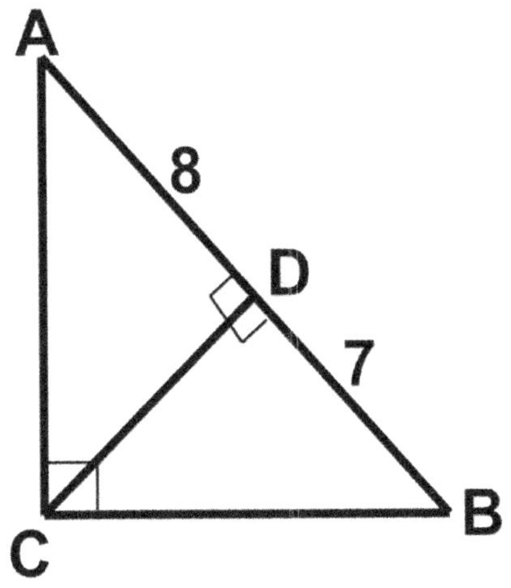


If $\overline{CD}$ is the altitude of $\triangle ABC$ then what is the length of $\overline{AC}$?

A) $2\sqrt{15}$

B) $2\sqrt{30}$

C) $\frac{15\sqrt{2}}{2}$

D) $\frac{15\sqrt{3}}{2}$

</td>
<td>

$\triangle ABC$ is similar to $\triangle ACD$ by AA because $\angle A \cong \angle A$ and $\angle ACB \cong \angle ADC$. Therefore, corresponding sides are proportional:

$\frac{AC}{AD} = \frac{AB}{AC}$ $\frac{AC}{8} = \frac{15}{AC}$ $AC = \sqrt{120} = 2\sqrt{30}$

SOLUTION: B

</td>
</tr>
</table>

**Comments on some official problems:**

2.1.25 The tangents of congruent angles are equal to each other, so find the tangent of angle T.

2.2.18 In this problem you are given extraneous information! The ratio of the sides is irrelevant, just find the angle that corresponds to angle T.

3.1.17 Angle F and angle J are congruent because they are corresponding angles of similar triangles.

4.1.9 Angle Q and Angle T are congruent.

## 4.10 Factoring $ax^2+bx+c$ when $|a| \neq 1$

Although students spend loads of time learning to factor a quadratic, it is a skill that is sometimes forgotten. If the leading coefficient is equal to $\pm 1$, a guess-and-check approach is fine. Otherwise, I suggest using the $ac$ method, as illustrated in two examples below.

**Example:** $4x^2-17x-15$

|  |
| --- |
| Place the $b$ coefficient, -17, into the top portion of the diamond. |
| Place the $ac$ product, -60 into the bottom portion of the diamond. Write the prime factorization of 60 as $60=2^2\cdot 3\cdot 5$ . |
| Find two factors of 60 that multiply to -60 but sum to -17. They are -20 and 3. |
| Take the original $4x^2-17x-15$<br>Use the factors to re-write the middle term<br>$(4x^2-20x)+(3x-15)$<br>Pull out common factors<br>$4x(x-5)+3(x-5)$<br>Factor by grouping $(4x+3)(x-5)$ |

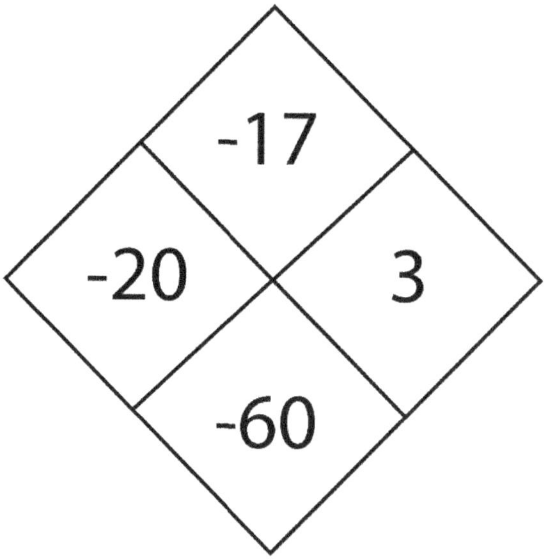


**Example:** $-10x^2-19x+2$

|  |
| --- |
| Re-write as $-(10x^2+19x-2)$ |
| Place the $b$ coefficient, 19, into the top portion of the diamond. |
| Place the $ac$ product, -20 into the bottom portion of the diamond. Write the prime factorization of 20 as $20=2^2\cdot 5$ . |
| Find two factors of 20 that multiply to -20 but sum to 19. They are 20 and -1. |
| Take the expression $-(10x^2+19x-2)$<br>Use the factors to re-write the middle term<br>$-[(10x^2+20x)-(x+2)]$<br>Pull out common factors<br>$-[(10x(x-2)-1(x-2)]$<br>Factor by grouping $-[(10x-1)(x+2)]$ |

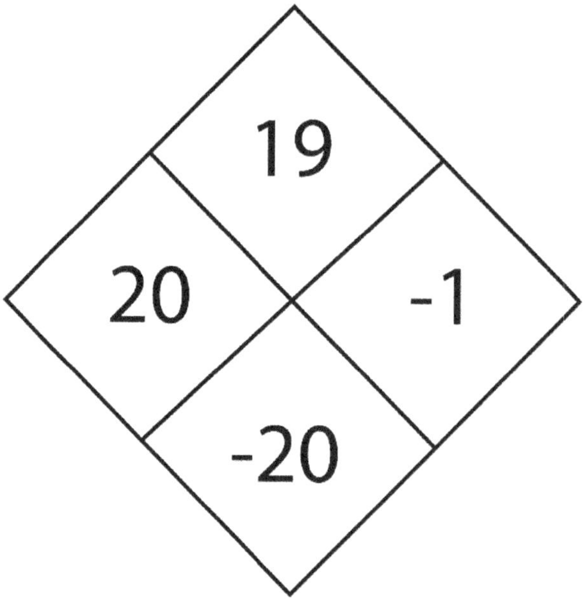


**Comments on some official problems:**

3.1.18 An interim step is to factor $2x^2+11x-546=(2x+39)(x-14)$

4.2.12 Set the equation to zero $4x^2+7x-36=0$ and factor $(4x-9)(x+4)=0$

Made in the USA
Las Vegas, NV
25 April 2023

71077568R00044